Modern Library Chronicles

ALSO BY EDWARD J. LARSON

EVOLUTION

EDWARD J. LARSON

EVOLUTION

*The Remarkable History of a
Scientific Theory*

A MODERN LIBRARY CHRONICLES BOOK

THE MODERN LIBRARY

NEW YORK

LIBRARY OF CONGRESS CATALOGING-IN-PUBLICATION DATA

Larson, Edward J. (Edward John)
Evolution : the history of a scientific theory / Edward J. Larson.
p. cm.—(A Modern Library chronicles book; 17)
Includes bibliographical references and index.
ISBN 0-679-64288-9 (alk. paper)
1. Evolution (Biology)—Philosophy—History.
I. Title.   II. Modern Library chronicles; 17.

QH361.L27 2004
576.8—dc22   2003064888

Modern Library website address: www.modernlibrary.com

Printed in the United States of America on acid-free paper

6 8 9 7 5

To

ERNST MAYR

IN COMMEMORATION OF HIS

100TH BIRTHDAY

JULY 5, 2004

A REMARKABLE LIFE IN SERVICE OF SCIENCE

Contents

LIST OF ILLUSTRATIONS

PREFACE

Nineteenth-century evolutionists envisioned the earth as a grand laboratory or workshop of organic development: a shimmering sphere of life spinning in a vast universe. That image inspired a new way of understanding nature. It changed how we view ourselves, one another, and all living things. We became interconnected competitors rather than separate creations. We now live in the shadow—or the illumination—of this modern biologic worldview.

The history of modern evolutionary science does not begin with Charles Darwin or even with biology. It begins with breakthroughs in late-eighteenth-century geology and paleontology. Indeed, when Darwin converted to an evolutionary view of biologic origins during the 1830s, he viewed himself as much as a geologist as a biologist. Fortunately, at the time, he did not have to categorize himself with either label, but instead could adopt the broader-brush title of "naturalist," which encompassed the various fields of scientific study that he drew upon in formulating his theory of evolution by natural selection.

Darwin's theory ripped through science and society, leaving little unchanged by its force. For nearly a century, scientists disagreed sharply among themselves over how evolution operates. Within the scientific community, a consensus answer to this question only began emerging during the 1930s, when a deeper understanding of genetics gave birth to the modern neo-Darwinian synthesis. Scientists still debate the details of evolutionary theory, however, and for many the devil lies in those details. Within the general population, disagreement continues even over whether species evolve, and most particularly over whether humans (or the essence of humanity) originated through purely natural processes from other forms of life. The stakes are enormous; few ideas more profoundly influence us than ideas about our origins. A starting point for any discussion of organic origins is an understanding of how the modern theory of evolution developed. It is a remarkable story of self-discovery that generated concepts affecting the very notion of what it means to be human. And it is far from finished. We will continue to learn more about organic origins—and about ourselves—for as long as we keep our minds open to new ideas in science.

This history of evolutionary science builds on my four previous books dealing with various aspects of the subject. I owe a continuing debt to the individuals and institutions that assisted me in my research and writing of those books, many of which are identified in them. To all of these people and places, thanks again. For this book in particular, I profited from the comments of Duncan Porter, Michael Arnold, Rodney Mauricio, and Thomas Lessl; the support of my wife, Lucy; and the enthusiasm of my children, Sarah and Luke. Thank you all. Finally, I am deeply grateful to Will Murphy of Random House. His vision brought this book into being; his editing made it better.

EVOLUTION

CHAPTER 1

BURSTING THE LIMITS OF TIME

Georges Cuvier had a large head—a famously large head—and an ego more than sufficient to swell even it. From his position atop the French scientific establishment during the first third of the nineteenth century, he accumulated high academic posts and official honors like some favored children collect toys: never enough and all kept in play. For his contributions to laying the foundations of modern biology, Cuvier willingly suffered comparisons to Aristotle, the acknowledged founder of the science. As a naturalist, Cuvier fancied himself "the French Newton"—bringing order to the life sciences much as Isaac Newton brought order to the physical sciences. Cuvier's rigorous empirical methods opened windows into the earth's biological history that would lead others to a vision of organic evolution he steadfastly refused to see. More than any other naturalist, he so greatly influenced the style and substance of nineteenth-century biology that the history of the modern scientific theory of evolution rightly begins with him—its staunchest foe.

Born in 1769 into an educated, bourgeois family in the Protestant, French-speaking portion of the independent French-German duchy of Württemberg, Cuvier was trained at a regional academy to serve in the duke's government. Pushed by his mother to excel academically, Cuvier's formal education included a solid introduction to natural history, a traditional subject encompassing such modern fields as biology, geology, oceanography, mineralogy, and paleontology. This subject became his passion. In 1788, with no government position open to him at home, Cuvier accepted employment as a private tutor for a French noble family in Normandy. There, as a sideline, he immersed himself in the study of marine invertebrates. From the relative safety of rural Normandy, Cuvier witnessed the French Revolution that began, from his perspective, with high hopes in 1789 but turned terribly ugly during the early 1790s. Becoming a citizen of France in 1793, when the French government annexed his homeland, Cuvier accepted a post in the revolutionary administration of Normandy even as he turned viscerally against the central regime's Terror and focused his own attentions on zoological fieldwork. In 1795, when a moderate republican government took power in Paris and promised to rebuild the central scientific establishment decapitated during the Terror, Cuvier moved to the capital in search of a career in science. There were plenty of openings for a naturalist of his obvious brilliance and driving ambition. Cuvier gained an assistantship at the renowned Museum of Natural History, and never looked back. His subsequent rise was meteoric. The study of natural history would never be the same.

Cuvier concentrated his scientific research on the burgeoning field of comparative anatomy; he was convinced that the internal structure of an animal revealed its function and therefore its true nature. In biology as in all else, form followed function for Cuvier. His research profited greatly from his position at the world's premier natural-history museum—an institution that rapidly became ever more comprehensive in its zoological holdings as Napoleon's armies plundered the collections of Europe and sent home live, preserved, and fossilized specimens from as far afield as Russia and Egypt. Ultimately, Cuvier proposed that there are four (but only four) basic anatomical types (he called them "*embranchements*") of animals: vertebrates (with backbones), molluscs (with shells), articulates (such as insects), and radiates (such as starfish). "Lesser divisions," he wrote, "are only modifications superficially founded on development or on the addition of certain parts, but which in no way change the essence of the plan."[1] This view, built solidly on anatomical analysis and still reflected (with modifications) in modern taxonomy, shattered the hierarchical concept dating from Aristotle of a single great chain of beings rising in fine gradations from the simplest living form to humans at its top. The idea within biology of seeing an anthropomorphic order in all living things gave way to studying them on their own terms.

Cuvier was the first naturalist to have at his disposal a suitably complete collection of the world's mammals—past and present—to make definitive distinctions among them. He made the most of this advantage, hoarding it to himself, his collaborators, and his protegés. In 1796, for example, he announced that, based on his anatomical comparisons of actual specimens, the elephants of India and Africa constituted two distinct species, and that both of them differed from the elephant-like mammoth found only in fossil remains. The positive identification of other living and extinct mammals followed one after another in rapid succession. To account for so many extinct species, as early as 1796 Cuvier announced "the existence of a world previous to ours, destroyed by some kind of catastrophe."[2]

Before Cuvier, European naturalists typically held that no species—all of them perfect in their original creation—ever died out. Fossils had no fundamental significance: Such things were simply sports of nature or remnants of some still-living species. Overturning this view, Cuvier ultimately concluded that *all* fossilized animals differed in kind from modern ones and that no modern species existed in truly fossil form. He boldly claimed the power "to burst the limits of time, and, by some observations [of fossils], to recover the history of the world, and the succession of events that preceded the birth of the human species."[3]

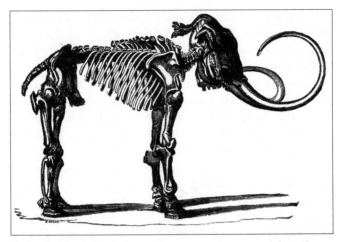

Early sketch of the skeleton of a mammoth extracted from frozen earth in Siberia, similar to the kind reported on by Georges Cuvier in 1796.

Suddenly, life had a history different from the present, and fossil fragments revealed it. "As a new species of antiquarian," Cuvier explained, "I have had...to reconstruct the ancient beings to which these fragments belonged; to reproduce them in their proportions and characters; and finally to compare them to those that live today."[4] The modern science of paleontology was born in Cuvier's laboratory. Because of his conviction that the form of any animal precisely served its functional needs, Cuvier confidently assumed that trained researchers could, in principle, reconstruct its entire structure from any one of its functional parts. Paleontologists could do for extinct animals what comparative anatomists did for living ones—definitively identify them. Doing so for all of the earth's past and present species became Cuvier's goal for science—and he himself would launch the effort, doing his own best work with fishes and four-footed mammals.

A compulsive worker, stern and impatient, Cuvier never doubted his own ability as a science researcher, educator, and administrator. He mastered the treacherous shoals of French academic politics just as ably as he mastered comparative anatomy. Even as he climbed the professional ladder within the Museum of Natural History, Cuvier gained leadership posts at the National Institute and the University of France— giving him unparalleled influence over patronage within the country's highly centralized science establishment. Napoleon named Cuvier to the Council of State in 1813, and he deftly kept his seat (and steadily expanded his portfolio) under three succeeding monarchs. Remarkably, even though every ruler he served was forcibly driven from office at least once, Cuvier held each of his official posts for life and died peacefully in his bed in 1832.

Napoleon ennobled him as a *chevalier;* Louis XVIII promoted him to the rank of baron; under Charles X, he became a grand officer of the Legion of Honor; Louis-Philippe made him a peer of France. "Cuvier was short and during the Revolution was thin," one biographer wryly noted. "He became stouter during the Empire; and he grew enormously fat after the Restoration."[5] Still there was that massive head, crowned with a thick mane of hair. According to one observer, Cuvier's head "gave to his entire person an undeniable cachet of majesty and to his face an expression of profound meditation."[6] Here was the lion of nineteenth-century French science and founder of modern comparative anatomy and paleontology. Yet his reasoned scientific arguments for the theory of special creation held back the tide of evolutionary thought, which had been rising since the Enlightenment, for a generation.

On the matter of organic evolution (or "the transmutation of species," as the concept was then called), it was not simply that Cuvier died before the publication of Charles Darwin's

Origin of Species and therefore never seriously considered the idea. He studied it carefully (albeit not in the light of Darwin's later arguments for it) and found it wanting. Although Cuvier's conclusions on this score reflected his religious and social beliefs, they were founded on his scientific understanding of nature. These added factors—religious and social—reveal telling aspects of pre-Darwinian Western thought about biological origins. They will be examined first.

Living in a particularly volatile era of French religious history characterized by alternating phases of Enlightenment scepticism, Revolutionary atheism, and Restoration Catholicism, Cuvier stood apart from most others within the cultural elite of France by remaining a churchgoing Protestant during his entire life. Indeed, he visibly aligned himself with his religious minority by overseeing government programs for Protestant education and serving as vice president of the Protestant Bible Society of Paris. He married a socially prominent Roman Catholic widow of the Terror, Anne Marie Coquet du Trazail, but they raised their children as Protestants. When his daughter Clémentine adopted an evangelical form of Protestantism, however, she grew to doubt her father's salvation and prayed for his conversion. That was not about to happen, at least on her terms. By definition, evangelicals publicly proclaim their religious beliefs and seek to convert others. But for Georges Cuvier religion was a strictly private matter. Perhaps it had to be so for him to prosper in French science and politics, but that lends an unjustifiably cynical slant to Cuvier's case. Although he was the very embodiment of reason in science, Cuvier accepted religious truth as existing wholly apart from reason. This made his private religious beliefs virtually invisible to others; despite considerable speculation, they have remained so to this day. Yet surely he was a Bible-believing Christian of some sort, and biblical Christianity carries with it certain presuppositions about origins. These presuppositions informed Cuvier's thinking about evolution just as they would for so many other Christians.

———

The biblical account of creation appears in the book of Genesis, which is sacred scripture for Jews, Christians, and Moslems. For the orthodox, Genesis represents the revealed word of God and, as such, carries special meaning in some literal, allegorical, or mystical sense. Even for liberal theists, led during the nineteenth century by a growing number of German and French theologians whose work Cuvier read, the Genesis account carries meaning as an early record of the Jewish people's understanding of God's role in Creation. Indeed, for those like Cuvier who accepted Moses as its author, Genesis gains authority as one of the earliest written records of creation. Accorded any of these meanings, the Genesis account becomes foundational for one's understanding of nature.

The first chapter of Genesis tells of God creating the heavens and the earth, then plants and animals, and finally humans—everything in six days. All types of plants and animals are said to reproduce "according to their kind." Read literally, this precludes evolution from one "kind" of plant or animal to another. Regarding humans, the account declares that God separately created them in His own image and likeness. The second chapter of Genesis contains an alternative creation account in which the order of the appearance of life forms on earth is somewhat reversed—but with a similar emphasis on the special creation of humans by God. Indeed, it is this second account that first introduces Adam and Eve as the progenitors of the human race, with God directly forming them as man and woman. The Bible does not state when these creation events occurred, but most early Christians probably assumed they all happened within the past six thousand years.

During the mid-1600s, Anglican archbishop James Ussher of Dublin used internal evidence within the Bible to calculate the year of creation as 4004 B.C., or less than three thousand years before Genesis was supposedly written by the Hebrew leader Moses. Printed in the margins of the Authorized, or King James, version of the Bible, Ussher's chronology became quasi gospel for British and American Protestants during the eighteenth and nineteenth centuries.

Generally speaking, Christian leaders from the early church through the Reformation did not view the Bible as a scientific text. They interpreted it as a divinely authored or inspired volume of separately written books and letters filled with spiritual meanings, some of them allegorical. Science, in the sense of a distinct intellectual tradition seeking rational explanations for physical phenomena, began with ancient Greek natural philosophy roughly five hundred years before Christ. Although most individual Greeks probably accepted religious or mythical explanations for natural phenomena, some Greek philosophers sought to separate the supernatural from the natural by proposing purely materialistic accounts of nature. Nothing is aught but physical matter in meaningless motion, the ancient Greek atomists proclaimed. The origin of life and individual species posed a particular problem for Greeks intent on devising purely materialistic explanations for natural phenomena. Creation implies a creator, and so to dispense with the need for a biological creator, such ancient philosophers as Anaximander, Empedocles, the atomists, and the Epicureans advanced various crude notions of organic evolution.

Based on his close study of animal anatomy, however, Aristotle concluded that species are absolutely immutable. Each species always breeds true to its form, he maintained, and never gives birth to a new type. Rejecting both creation and evolution, Aristotle (an atheist) simply posited that species are eternal. Integrating the Genesis account with mainstream Aristotelean science, premodern Christian naturalists viewed species as created by God in the beginning and thereafter fixed for all time in a perfect (albeit fallen) creation. Well into the nineteenth century even Cuvier saw no scientific reason to reject Aristotelean thinking on the fixity of species—and fully appreciated the religious advantages of retaining it. Whether read literally or allegorically, the Genesis account harmonizes with the idea that species, once created, never change.

———

By 1800, Cuvier also had compelling social reasons for maintaining the traditional Aristotelean view of species. The breakdown of established authority associated with the Enlightenment in eighteenth-century France coincided with a revival of pre-Christian speculation about organic evolution and biological reductionism or materialism. At its core, the Enlightenment (the intellectual launching pad of modernity) involved a rational critique of previously accepted doctrines and institutions. To the extent that Christianity was based on divine revelation rather than human reason, it lost credibility among enlightened thinkers. Similarly, to the extent that they lacked rational justification, political and cultural institutions trembled or fell—including the ancien régime. During the 1790s in France, King Louis XVI lost his head and the Roman Catholic church was outlawed. Revolutionary currents swirled through natural history, as well. Some radical naturalist and savants challenged static concepts in science, including the fixity of species; many of them rejected any ongoing role for the supernatural in the natural. Rational materialism gained ground in scientific, social, and political thought—with no clear separation among these disciplines. Disorder became the order of the day, and a reaction became inevitable.

Among eighteenth-century French scientists, Georges-

Louis Leclerc, comte de Buffon, personified the Enlightenment. One of the foremost descriptive naturalists of his day,
Buffon was also a highly original theorist who, as superinten-
dent from 1739 to 1788 of the Royal Garden (which became
the Museum of Natural History after the Revolution), com-
manded the position and prestige to promote his novel ideas
about nature. In all these respects—academic field, official
status, and scientific renown—Buffon was Cuvier's predecessor but never his precursor. Although historians still debate
whether Buffon was an outright atheist or simply a radical
deist, he certainly rejected Christianity and sought materialistic explanations for the origin of the earth and its inhabitants. This led him to evolutionary thought.

Scientific materialism ran through Buffon's massive treatise, *Natural History*, which appeared in fifteen initial and
seven supplemental volumes over the forty-year period from
1749 to 1789. The earth and other planets congealed from
globs of molten matter thrown off when a comet crashed into
the sun, Buffon proposed in Volume One, and living things
spontaneously generated on the earth as it cooled, he added
in later volumes. For proof, he offered crude experiments
with molten iron balls, whose cooled surfaces conveniently
wrinkled like the earth's terrestrial surface, and boiled meat
gravy, which became alive with microorganisms when cooled.
By Volume Fourteen of his treatise, Buffon was speculating
about the evolutionary origins of similar species from common ancestral types—perhaps as few as thirty-eight original
forms for the two-hundred-odd mammalian species known at
the time, he estimated. For example, he proposed that all the
world's various lions, tigers, leopards, pumas, and domestic
cats "degenerated" in response to local climatic conditions
from a single ancestral type of cat. This constitutes evolution,
at least on a limited scale. As evidence, Buffon offered the observation that native American mammals (be they cats, bears,

or people) were invariably smaller and weaker than their Old World counterparts: Surely the American types had degenerated due to the New World's harsh climate. Such thinking pushed God as creator either back in time or out of the picture altogether—Buffon never made it clear which. It also enraged Thomas Jefferson, who countered with depictions of large, strong native American animals in his 1787 book, *Notes on the State of Virginia.*

Buffon placed limits on his materialism by postulating that certain "internal molds" guide the spontaneous generation and subsequent degeneration of living organisms. Due to these molds, each basic biological family remains distinct through time, he asserted. One type of cat might degenerate into another type of cat, but never into a dog. Buffon did not say who or what designed these internal molds, but their ongoing existence retained a foundational element of design in nature. Thus, unlike under some later theories of evolution, neither the original generation of nor subsequent variations in living things was utterly random. In fact, in the second supplemental volume of his *Natural History,* Buffon asserted that, under similar climatic conditions, fundamentally similar species would spontaneously generate on any planet. For him, internal molds were universal and eternal—nothing less could explain the apparent orderliness of life; anything more might leave too much room for God.

Any metaphysics was too much for the extreme materialists of the French Enlightenment. The radical mid-eighteenth-century encyclopedist Denis Diderot saw no trace of design in nature, for example. Any form of living thing could spontaneously generate itself in a purely material process, he argued. Those forms that could survive and reproduce would do so while all others would die out, leaving the current diversity of life on earth, without any divine intervention along the way. Although Diderot envisioned a purely random process spawning and modifying life on earth (much as orthodox Darwinists later maintained), he (unlike them) attributed a primitive awareness to the matter that self-generated into living beings. Paul Henri Thiry, baron d'Holbach, dispensed with even this bit of retained vitalism. In his 1770 *System of Nature,* widely known as "the Bible of atheism," Holbach simply asserted that inert matter could self-organize into complex structures; when they became complex enough, these material structures would exhibit the properties of life. Though Diderot and Holbach drew on the work of various naturalists (including Buffon) to support their philosophical speculations, neither conducted original scientific research. As prophets of atheism and champions of the radical Left, they mainly sought to account for the origin of life without recourse to design or a designer, and paid scant attention to technical points of how species are formed. As a consequence, their writings had greater philosophical than scientific impact.

Some Enlightenment naturalists found evidence of materialism at the foundation of life, however. In 1740, Swiss naturalist Abraham Trembley discovered the "polyp" or freshwater hydra—a peculiar little animal that could regenerate multiple complete beings when cut into parts. A fellow Swiss naturalist, Charles Bonnet, soon found that some species of worms could do the same. Meanwhile, Swiss physician Albrecht von Haller demonstrated that certain animal tissues, including human muscles, reacted directly to electrical shock without any intervention of the brain or a soul. Inspired by these observations, French naturalist Julien Offroy de la Mettrie described life as a basic principle of organic matter itself, and not as the product of an independent mind or indwelling soul. Such thinking reached into popular culture. Drawing on scientific speculation about subtle electrical and magnetic fluids animating life, for example, during the years leading up to the French Revolution, German physician Franz Mesmer established a highly lucrative practice treating the muscle cramps and headaches of wealthy Parisians (including Queen Marie-Antoinette) with magnetic cures featuring an early form of hypnosis that became known as "mesmerism."

Of course, no one can precisely gauge how much Enlightenment thinking about material origins for life and organic species contributed to the political, social, and religious turmoil of the French Revolution, but some observers at the time saw a causal link. Notions of biological instability seemed to breed social disorder; rational materialism undermined traditional political and religious authority; and chaos ensued as the law of the jungle became the rule in Paris. For Cuvier, who was traumatized by the Revolution and thereafter sought political and social order above all, this connection damned the very idea of organic evolution and all manners of biological speculation. Empirical facts alone could provide a solid basis for science and society, he believed. Speculative systems in natural history invited ruin. "Persuaded as I am of the futility of all these systems, I find myself pleased each time a well-established fact comes and destroys one of them," Cuvier proclaimed in 1804.[7] He reported to Napoleon four years later, "Our natural sciences are only the facts brought together, our theories only formulae which embrace a great number of them."[8]

Cuvier took every opportunity to criticize the speculations of Buffon and other materialist-minded naturalists. In his 1796 inaugural lecture at the National Institute, he made a point of attacking Buffon's theory of evolutionary degeneration even though Buffon was long dead and his theory had not attracted many followers. Pointing to functional anatomical differences between Indian and African elephants, Cuvier declared, "Whatever may be the influence of climate to make animals vary, it surely does not extend this far. To say that it can change all the proportions of the bony framework, and the intimate texture of the teeth, would be to claim that all quadrupeds could have been derived from a single species; that the differences they show are only successive degenerations; in a word, it would be to reduce the whole of natural history to nothing, for its objects would consist only of variable forms and fleeting types."[9] From this initial outburst to his final public debates with the evolutionary naturalist Étienne Geoffroy Saint-Hilaire in 1830, Cuvier opposed the idea of organic evolution in all its forms. At the time, he was not alone in seeing a radical tint to the so-called "transmutation hypothesis."

———

However much his religious and social views may have biased him, Cuvier had solid scientific reasons for rejecting the concept of organic evolution. Those reasons were not reactionary. Quite to the contrary; they reflected the most progressive science of the day—science that, in Cuvier's own words, burst the traditional limits of geologic time. His findings in both primary fields of his scientific research, comparative anatomy and paleontology, convinced Cuvier that evolution was impossible.

Working under Buffon, Louis Jean Marie Daubenton pioneered the study of comparative anatomy at the old Royal Garden in Paris, focusing his work on the external characteristics and major organs of animals. Shifting his research to mineralogy after the Garden's post-Revolution reorganization into the Museum of Natural History, Daubenton supported the appointment of Cuvier to assist with anatomical studies. Building on the work of Daubenton's former assistant, Félix Vicq d'Azyr, who died during the Revolution, Cuvier stressed the importance of examining an animal's entire internal structure down to its smallest parts. Close analysis of mammals gave Cuvier an unprecedented appreciation of what he came to see as the irreducible functional complexity of living things. "Today comparative anatomy has reached such a point of perfection that, after inspecting a single bone, one can often determine the class, and sometimes even the genus, of the animal to which it belongs," Cuvier commented in 1798. "This is because the number, direction, and shape of the bones that compose each part of an animal's body are always in a necessary relation to all the other parts, in such a way that—up to a point—one can infer the whole from any one of them."

Here he added a telling example. "If an animal's teeth are such as they must be in order for it to nourish itself with flesh, we can be sure without further examination that the whole system of its digestive organs is appropriate for that kind of food; and that its whole skeleton and locomotive organs, and even its sense organs, are arranged in such a way as to make it skillful at pursuing and catching its prey."[10]

From his study of animal structure, Cuvier devised his doctrine of the correlation of parts. "Every organized being forms a whole, a unique and closed system, in which all the parts correspond mutually, and contribute to the same definitive action by a reciprocal reaction," he wrote in the widely read "Preliminary Discourse" to his landmark 1812 *Recherches sur les Ossements Fossiles.* "None of its parts can change without the others changing too; and consequently each of them, taken separately, indicates and gives all the others."[11] Taken seriously (and Cuvier took everything he wrote with the utmost seriousness), this doctrine precludes organic evolution. No materialistic evolutionist ever claimed that *all* an animal's parts changed simultaneously—that would be as grand a miracle as divine creation. For evolution to occur, the process must involve an accumulation of changes. Yet Cuvier concluded, based on his research, that anatomical interactions

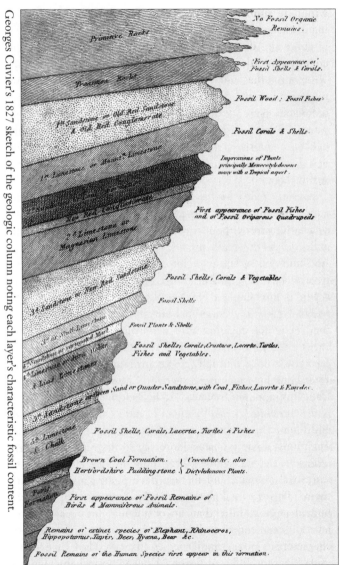

Georges Cuvier's 1827 sketch of the geologic column noting each layer's characteristic fossil content.

within an animal are so delicately balanced that any mean-
ingful change in one would render the whole being incapable
of survival. Add to this doctrine a basic appreciation of the
complex interdependence among species and the common
observation that individuals of a species appear to breed true
to type, all of which Cuvier did, and the conclusion becomes
inevitable. New species cannot evolve from old ones. In the
wake of Buffon, some radical French naturalists continued to
debate the merits of transmutation despite Cuvier's doctrine
of the correlation of parts, but the weight of its authority
strongly reinforced the creationist position.

   To his anatomical arguments against evolution, Cuvier
added paleontological ones. At the time, no one knew the fossil record better than Cuvier. His museum housed the finest
collection of fossils in the world and his collaboration with
Alexandre Brongniart in fieldwork around Paris laid bare successive layers of geologic strata, each with its distinctive fossil
types. Before Cuvier, few people found many fossils anywhere. He found them everywhere and gave them new meaning. "Whether one digs into the plains, or penetrates into
caves in the mountains, or climbs their torn flanks, one encounters everywhere the remains of organisms," he noted in
1800. "Immense masses of shells are found at great distances
from any sea [and] seams of coal display the imprints of
plants at heights or depths that are equally striking. But what
is still more surprising is the disorder that reigns in the accu-
mulations of these objects: here, shelly beds are covered by
others that contain only plants; there, fish are superposed to
terrestrial animals, and in turn have plants or shells above
them." This was (and is) the geologic column, and Cuvier exhumed it more fully than anyone before him: Layer upon
layer of sedimentary rock, with each layer containing its own
characteristic mix of fossil species, like a wedding cake whose
every tier is a different flavor. "Almost everywhere," Cuvier

added, as if to comment on the cake's misfit icing, "these remains of organisms are utterly foreign to the climate of the ground that conceals them."[12]

For Cuvier, the geologic column suggested a historical pattern of catastrophic floods—some global, others local—alternating with periods of terrestrial uplift. These were massive deluges, he surmised, such as the world has not seen at least since the biblical time of Noah. Each of these floods drowned all terrestrial life within its reach, scoured the terrain, and laid down a layer of sedimentary rock containing the fossil remains of the former population. When the land resurfaced in a later geologic epoch, different kinds of plants and animals repopulated it—perhaps through migration from other regions, perhaps by their divine creation. Cuvier's writings suggested both these options, but never committed him to either. The scientific evidence is inconclusive as to when the various species originated, Cuvier opined, so he would leave the matter to speculation (rather than science). At the most, he wrote, "One is authorized to believe that there has been a certain succession in the forms of living beings."[13]

Although Cuvier did not know when individual species originated, he was certain that some of them had become extinct, which was news enough in 1800. This discovery came from establishing beyond a reasonable doubt that fossilized animals found in the older layers of the geologic column differ in kind from living ones. "It is the generality of this difference that makes it the most remarkable and astonishing result that I have obtained from my research," Cuvier proclaimed at the dawn of the new century. "I can now almost assert that none of the truly fossil quadrupeds that I have been able to compare precisely has been found to be similar to any of those alive today."[14] As his investigations progressed, Cuvier further generalized this assertion to cover various types of animals from different geologic epochs. For example, he later reported that the types of marine molluscs found in any particular layer of the geologic column appeared only in that layer—never in earlier or later ones.[15]

The geologic column did reveal a progression of forms from mostly simple ones toward the bottom to more complex ones toward the top, Cuvier observed. The types of animals also became increasingly like present-day ones over time. For example, he noted that marine molluscs from recent fossil beds are more like living molluscs than those found in ancient beds. To some, such evidence suggested evolution. But Cuvier had already rejected this explanation based on his study of comparative anatomy, and the apparent absence of transitional forms in the fossil record confirmed this conclusion. Certainly he saw a succession of comparable types over time but, to his eye, the series progressed by jumps rather than gradations. Indeed, in all his extensive study of fossils, Cuvier perceived only distinct species that persisted without change, never a gradual blurring of one type into another.[16] Of course, his doctrine of the correlation of parts held that such change was impossible anyway.

The history of current species provided further evidence for Cuvier's anti-evolutionism. The animals depicted in ancient drawings and the mummified ibises in Egyptian tombs appear identical to their living descendants, he noted. "I am well aware that there I am citing monuments of only two or three thousand years ago," Cuvier conceded, "but that is as far back as it is possible to reach."[17] It should stand for something. For Cuvier, every known scientific "fact" pointed toward the fixity of species.

———

Despite his conservatism, Cuvier was a visionary figure in paleontology. "There is," he declared, "a series of epochs anterior to the present one, the order of which can be verified without uncertainty, although the duration of the intervals between them cannot be defined with precision." Cuvier's earth was very old—far older than Genesis suggested. He saw evidence of a vast period extending before the appearance of any living organisms followed by multiple geologic epochs, each populated with its own distinctive flora and fauna. Relics of all current species, including humans, appear only in the most recent geologic deposits, he asserted, and never among truly fossilized types. "Thus life on earth has often been disturbed by terrible events," Cuvier concluded. "Living organisms without number have been the victims of these catastrophes. Some were destroyed by deluges, others were left dry when the seabed was suddenly raised; their races are even finished forever, and all they leave in the world is some debris that is hardly recognizable to the naturalist."[18] Cuvier developed his theories of catastrophes and migration from the ideas of earlier naturalists but, as his biographer William Coleman notes, one feature of Cuvier's geologic doctrine stood out as breathtakingly novel: "the demonstration by comparative anatomy of the fact of organic succession."[19] Cuvier's protests notwithstanding, this demonstration proved foundational for modern evolutionary thought.

In his basic conception of natural history, Cuvier carried mainstream scientific opinion with him. In fact, his rhetorical invocation of "facts" and repudiation of theorizing particularly appealed to conventional, conservative naturalists—precisely those scientists most likely to accept traditional viewpoints, including biblical ones. After a generation of battering by radical French materialists, some Christian intellectuals (particularly in Britain and the United States) welcomed the findings of a prominent French naturalist whose views were not openly hostile to their own. They met him at least halfway. In 1813, a year after the French publication of Cuvier's popular "Preliminary Discourse," the pious Scottish geologist Robert Jameson translated it into English under the

title *Essay on the Theory of the Earth,* with a preface and notes
stressing points where the French naturalist's views coincided
with Christian doctrine. Working in the shadow of French rationalism, Cuvier never invoked scriptural authority to support his scientific arguments, but his "Preliminary Discourse"
did observe that the date traditionally ascribed to the biblical
deluge that supposedly drowned all life outside Noah's ark
roughly coincided with geologic evidence for the time of the
last catastrophic flood.[20] Jameson's annotations trumpeted this
observation and Cuvier's devastating critique of evolution.[21]

During the first half of the nineteenth century, other conservative Christian geologists in Britain and America labored
to reconcile the new scientific orthodoxy with the Genesis
account. In 1814, Scottish natural theologian Thomas
Chalmers proposed that a gap existed in the Genesis narrative between the book's first and second verses. This opened
unlimited time for geologic epochs between "the beginning"
and God's creation of current species. Amherst College geologist Edward Hitchcock adopted this so-called "gap theory"
and popularized it in the United States. Meanwhile, Scottish
geologist Hugh Miller suggested that the days of creation in
Genesis symbolized geologic epochs. Yale University geologists Benjamin Silliman and James Dwight Dana (a father-in-
law, son-in-law team) championed the "day-age theory" in
the United States.

During the mid–nineteenth century, Cuvier's followers
modified his basic outline of geologic history to keep abreast
of the latest scientific findings. For example, to avoid choosing
between a single creation of all life and multiple creations
following the various catastrophes, Cuvier maintained that
migration could account for the abrupt appearance of new
species in the local fossil record of a particular place. As
wider fieldwork gradually eliminated plausible sources for
the migrating species, however, many of Cuvier's followers

turned to multiple creations as the most realistic explanation for the abrupt appearances of species in the fossil record under a creationist model. Cuvier's equation of the biblical flood with the final catastrophe lost its principal proponents in the 1830s, when British geologists Adam Sedgwick and William Buckland, both conservative Christians, concluded that a single deluge of the type described in Genesis could not produce the complex deposits attributed to the last catastrophe. For them, the biblical flood lost all geologic significance. By that time, Cuvier's Swiss disciple, Louis Agassiz, had shown that ice ages (rather than floods) probably caused the catastrophic extinctions recorded in the fossil record. With such modifications, Cuvier's creationism remained the dominant scientific explanation for the origin of species until the 1860s, when it vied with Darwin's theory of evolution for acceptance by mainstream scientists.

Cuvier's legendarily large head generated a telling anecdote. It seems that he often left his hat in his outer office. Some visiting scientists, waiting to meet the great man, could not resist the temptation to don it. The hat inevitably slipped over their ears and covered their eyes. Cuvier's scientific thought had somewhat the same effect on natural history. Broad and all-encompassing, it blocked alternative theories from sight—at least until a scientist of Darwin's stature lifted the brim.

CHAPTER 2

A GROWING SENSE OF PROGRESS

Cuvier may have given meaning to biologic history, but dinosaurs made it come to life. The recognition that a succession of species populated the earth was Cuvier's principal contribution to paleontology. He never envisioned the process as progressive, however, nor conceded the introduction of new species over time. Given the geographically limited range of his evidence, Cuvier could maintain (with decreasing credibility) that the fish-to-reptile-to-mammal-to-man sequence he uncovered could have resulted from the migration of the later-appearing species from areas whose fossils were not yet known. Any suggestion that the pattern reflected the evolution or progressive creation of increasingly complex animals over time would be sheer speculation, he asserted.

Overlapping with Cuvier's early (and most original) work on these matters, the English geologic cartographer William Smith found that he could definitively identify each stratum of sedimentary rock across Britain by the characteristic mix of the fossil species that it contained. Smith's findings reinforced and extended the concept of organic succession, but they, too, did not definitely demonstrate direction over time. Yet other naturalists soon made similar findings in fossil beds elsewhere, so that by the 1820s the conclusion became increasingly inescapable: New species both appeared and disappeared over time. There was simply no place where the newly appearing species could have migrated from: They must truly be new. With traditional faith in a single creation and the permanent endurance of species shattered, the search was on to discover more types of past life. Among the new-found fossil species, none attracted more attention than dinosaurs, particularly after their identification as gargantuan ancient reptiles. The fossil record appeared to show direction over time from a reptilian age to the present ascendancy of mammals.

It was in fact Cuvier who identified the first-known great reptile of a bygone era from a four-foot fossil jawbone captured by the French republican army during its sweep of the Netherlands's Meuse region in 1795. Known as *Mosasaurus* (or "lizard of Meuse"), this was a huge reptile from Secondary (or Mesozoic) chalk beds near Maastricht. These much-quarried ancient beds contained a rich array of fossil invertebrates, fish, and reptiles, but no mammals. Cuvier identified *Mosasaurus* as an extinct marine lizard with anatomic similarities to present-day monitor lizards of the tropics, but much bigger and solely aquatic. The honor of identifying the first great land lizards, later classified as dinosaurs, fell to an incongruous pair of English naturalists, William Buckland and Gideon Mantell.

WHEN GIANTS WALKED IN ENGLAND

Buckland was a boisterous polymath of the early Victorian era. Ordained an Anglican cleric and elected a fellow of Corpus Christi College, Oxford, in 1809, Buckland was tapped in 1818 for a new Oxford University readership in geology, a subject that he had loved since his childhood. He rose to the top of the British scientific establishment over the next three decades, serving twice as president of the Geological Society of London and once as president of the British Association for the Advancement of Science—all the while holding ecclesiastical posts ranging from a country rectory near Oxford to dean of Westminster. Despite his prestigious positions,

Buckland never took himself (or his colleagues) too seriously. His flamboyant lecturing style became legendary at Oxford. To illustrate his lectures on dinosaurs, for example, he might lumber around the lectern mimicking the gait of an over-stuffed land lizard or flap his clerical coat tails like a winged pterodactyl. Known for attributing the odd collection of pre-historic animal remains found in a Yorkshire cave to its sup-posed use in warmer pre–Diluvial times as a hyena den, Buckland kept a live African hyena at his home, along with a pet bear, who accompanied the geologist to college functions wearing a cap and gown. Spoofing his reputation for discover-ing fossil reptiles, he served alligator meat to favored guests. As a young man, Charles Darwin found Buckland's antics off-putting, and attributed them to "a craving for notoriety, which sometimes made him act like a buffoon"—but then, Darwin went to Cambridge.[1]

An avid collector of specimens, Buckland obtained his first dinosaur fossils from a slate quarry near Oxford sometime during the late 1810s. "The detached bones," he later wrote, "must belong to several individuals of various ages and sizes. . . . Whilst the vertebral column and extremities much resemble those of quadrupeds, the teeth show the creature to have been oviparous, and to have belonged to the order of Saurians or Lizards."[2] Their size was striking. One thighbone measured nearly three feet long and ten inches around. If its relative proportions matched those of living lizards, then this animal had "a length exceeding 40 feet and a bulk equal to that of an elephant seven feet high," Buckland reported.[3] It took some time for him to figure out what he had found, but it came from Secondary strata roughly the same age as the one that held Cuvier's *Mosasaurus* and, emboldened by that precedent, in 1824 Buckland finally published a description of his *Megalosaurus* (or "great lizard"), the largest land reptile heretofore identified.

By the time that Buckland published his discovery, Gideon Mantell had found even grander fossil remains of the same species in southern England. "The beast in question would have equaled in height our largest elephants, and in length fallen but little short of the largest whales," Buckland reported.[4] Mantell practiced surgery for a living, but loved nothing more than to collect fossils—making his best finds in a sandstone quarry near his home in Sussex. His *Megalosaurus* fossils came from this site in 1821. Mantell's wife, Mary Ann, often accompanied her husband on his expeditions, until she became so tired of his morose, sometimes paranoid, obsession with fossils that she left him. She was along in 1822 when one of them found in exhumed rock an enormous fossil tooth, worn down like those of a plant-eating mammal. Yet mammal fossils did not come from this Secondary strata. Even Cuvier was stymied by the tooth, at first identifying it as from a rhinoceros, and only later writing to Mantell, "Might we not have here a new animal, a herbivorous reptile?"[5] Mantell confirmed Cuvier's suggestion by comparing the tooth with those of present-day iguanas—finding a match in everything except size. His tooth would have come from a sixty-foot-long iguana, Mantell estimated. He published his find in 1825, making his *Iguanodon* (or "iguana tooth") the second dinosaur of record. Mantell identified a third species of dinosaur, the armor-plated *Hylaeosaurus* (or "woodland lizard"), eight years later from a remarkably complete fossil embedded in a block of Sussex limestone. In 1841, these three fossil species, by then confirmed in multiple specimens, became the founding members of *Dinosauria*, a newly minted prehistoric suborder of saurian reptiles.

Buckland and Mantell labored to fit their dinosaurs into a broad temporal context, and for both of them that context took on an increasingly progressive aura. They were Cuvierian catastrophists to be sure, but like many naturalists in the

1820s, they were sensing direction in biologic history. Cuvier had confined his analysis to two basic divisions of sedimentary rock—earlier Secondary strata, rich in fossilized fish and marine reptiles, and later Tertiary (or Cenozoic) strata, where the fossils of land mammals first appear in great numbers. He never presented this as a progression of types over time. If to anything other than migration, he attributed this biologic sequence to a hypothesized retreat of the primeval oceans, which would have exposed more land for mammals in Tertiary times. Although Cuvier ducked the issue, the discovery of large land reptiles in Secondary strata disrupted the simple equation of earlier times with aquatic animals: The presence of dinosaurs implied the existence of vast pre-Tertiary continents. In an 1831 paper, for example, Mantell placed his dinosaurs in a "geological age of reptiles," in which giant lizards ruled the sea, land, and sky. It followed an age of fish and preceded an age of mammals—a seemingly directional sequence.[6] Befitting his academic and ecclesiastical status, Buckland offered an even grander and more progressive-minded scheme than did Mantell.

By the early nineteenth century, European naturalists generally accepted the nebular hypothesis, first fully articulated by French astronomer Pierre-Simon Laplace in 1796, as a probable explanation for the origin of stars and solar systems. Our solar system began as a nebular cloud of rotating dust and gas that slowly collapsed under the weight of its own gravity, Laplace proposed. Most of the nebula's mass condensed into a dense, rotating core, so hot that it glowed: our sun. As the nebula collapsed, it left behind rings of rotating matter that coalesced into molten, spherical planets, some with their own rotating rings or coalesced moons. Here was evolution by natural law with a vengeance. As for God, Laplace reportedly quipped to Napoleon in 1802, "I have no need of that hypothesis."[7] The telescopic discovery of gas nebulas in space by the great English observational astronomer William Herschel provided persuasive evidence for Laplace's hypothesis, though Herschel and many others who accepted it never endorsed Laplace's ouster of God from a participatory role in the process. Under the nebular hypothesis, the earth's surface would have gradually cooled from its molten beginnings to its current temperate state. Some naturalists, including Buckland, saw this climatic change as a proximate physical cause for organic succession. A thoroughly Christian catastrophist, Buckland envisioned a good God creating a progressive succession of species, each perfectly designed for the climate of its particular geologic epoch and all pointing toward the ultimate creation of humans in God's image when conditions became right.

Of course, Buckland's view of biologic history drew on biblical revelation as well as paleontologic evidence, but not in a strictly literalistic manner. From the outset, Buckland posited a gap in the Genesis account that left ample time for a long geologic history before the creation of current forms, and he gradually recognized a diminished role for the biblical Deluge in shaping current geologic features. These liberal interpretations elicited bitter opposition from conservative Christians on the Oxford faculty and throughout England. Yet Buckland remained deeply committed to the core principles of natural theology, which saw evidence of God's existence and beneficence in nature. "Minds which have been long accustomed to date the origin of the universe, as well as that of the human race, from an era of about six thousand years ago, receive reluctantly any information, which if true, demands some new modification of their present ideas of cosmogony," he explained in 1836, "and, as in this respect, Geology has shared the fate of other infant sciences, in being for a while considered hostile to revealed religion; so like them, when fully understood, it will be found a potent and consistent auxiliary to it, exalting our conviction of the Power, and Wisdom, and Goodness of the Creator."[8]

Buckland was a throughly rational Christian. When encountering an alleged miracle of martyr's blood perpetually wetting the floor of a Roman Catholic cathedral, he tested the hypothesis by licking the spot with his tongue. "Bat urine," the Anglican cleric pronounced.[9] Buckland's God used systematic processes to guide terrestrial events with a designer's touch; his God did not intervene irrationally. For Buckland and many other early-nineteenth-century British naturalists, the succession of species in the fossil record reflected God's direction for life on earth. It had a beginning and we are its end. They never presumed to explain precisely how God created new species at the dawn of each epoch—as a divine act, it was beyond the realm of science.

———

In the same year that Buckland became Oxford's first reader in geology, 1818, Cambridge named Adam Sedgwick its geology professor. By his own perhaps overly modest admission, he knew nothing about the subject at the time, but was a quick study. He held the professorship for more than fifty years, becoming an institution at Cambridge and within British science. Kindred spirits, Buckland and Sedgwick set the norms for their field in England for a generation. But where Buckland did his best work on recent geologic features, Sedgwick sought to penetrate the fossil record back to the earliest vestiges of fossilized life on earth. This was virgin soil for geology in the 1820s, and Sedgwick found an ideal place to look for it in the ancient rocks of Wales. There Sedgwick discovered the Cambrian system (so named in 1835), the oldest strata of fossil-bearing rock, deep in the Transition (or Paleozoic) series underlying Cuvier's Secondary formations. Here trilobites reigned.

Summer after summer, Sedgwick worked the ancient

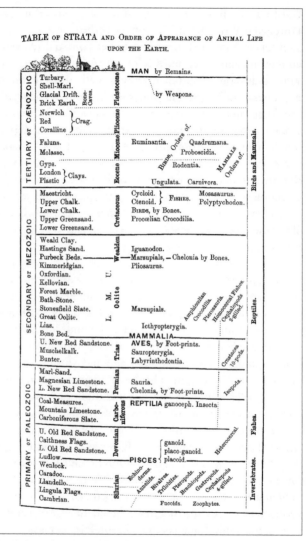

Richard Owen's 1861 table of geologic eras, periods, and epochs, with order of appearance of biologic types.

Welsh strata from the north while Sir Roderick Impey Murchison, a retired army officer who later led the government effort to survey Britain's geology, worked it from the south. They found fossilized fishes in the upper reaches of the Transition series (the Devonian and upper Silurian) but only the remains of long-lost invertebrates (like trilobites) in the Cambrian. Separate, well-defined ages of invertebrates and fishes appeared to precede the familiar ages of reptiles and mammals in the fossil record. In 1841, John Phillips (nephew and student of pioneer geologic cartographer William Smith) formally divided the geologic column temporally by bestowing new names on its layers. The old Transition series became the Paleozoic Era, an age of invertebrates and fishes; the middle Secondary series became the Mesozoic Era, an age of reptiles; and the young Tertiary series became the Cenozoic Era, an age of mammals. These eras were defined by sharp breaks in the fossil record, as were the various periods within each era (such as the Cambrian), but a trend line stood out. "Now, I allow (as all geologists must) a *kind* of *progressive development.* For example, the first fish are below the reptiles; and the first reptiles older than man," Sedgwick wrote in 1845. This directional finding simply reinforced Sedgwick's pious creationism, for he immediately added: "I say, we have successive forms of animal life adapted to successive conditions (so far, proving design), and not derived in natural succession in the ordinary way of generation" by transmutation or organic evolution. "How did they begin?" Sedgwick asked and answered. "I reply, by a way out of and above common known, material nature, and this way I call *creation.*"[10]

Sedgwick, Buckland, and other naturalists of their ilk saw a pattern of successive creation in the fossil record, with God as its active source and a cooling earth as its mechanical regulator. Acting in a discontinuous manner, God lovingly designed new populations perfectly fitting the ever-cooling,

ever-improving terrestrial climate while mercifully destroying the preceding populations when they no longer fit, leading to the creation of humans in what some saw as the
biblically prophesied end times. Reviewing the succession of
forms in the fossil record from his Cambrian invertebrates to
"the recent appearance of man," Sedgwick thus affirmed in
his 1831 Presidential Address to the Geological Society of
London, "There has been a progressive development of organic structure subservient to the purposes of life."[11]

EVOLUTION IN THE AIR

Even as catastrophe-minded naturalists refined their theories
of a successive creation fitted to an improving environment,
some less orthodox scientists saw evidence of organic evolution in the progression of fossil forms and living species. Indeed, Sedgwick's above-quoted 1845 outburst against
evolution was expressly directed at "the opinions of [French
naturalist] Étienne Geoffroy Saint-Hilaire and his dark
school [that] seem to be gaining some ground in England."[12]
Sedgwick probably referred here to two recently published
British books, Robert Grant's 1841 *Outline of Comparative
Anatomy* and Robert Chambers's 1844 *Vestiges of the Natural
History of Creation,* which explicitly enlisted the emerging fossil record to promote the idea of transmutation. Grant had
been openly flirting with evolutionism since the 1830s;
Chambers was a new convert who wrote anonymously.

As their common inspiration, Geoffroy, Grant, and Chambers looked to the writings of Jean Baptiste Pierre Antoine
de Monet, chevalier de Lamarck, Cuvier's colleague at the
French Museum of Natural History who in 1802 published
the first comprehensive theory of organic evolution—a
highly progressivist account known as "the transmutation hypothesis" or, later, simply as "Lamarckism." Using the evi-

dence available to him at the time, Cuvier had battered Lamarck's hypothesis until it was widely discredited and all but dead. But no scientific idea ever really dies; a bit of new evidence, and it can spring back to life. This happened to Lamarckism, beginning with the fossil discoveries of the 1820s and 1830s, and even more so with those of the 1880s and 1890s. Aspects of it still linger on the fringes of science, grasping for the evidence that might revive it once more.

Lamarck lived from 1744 to 1829. The eleventh child in a semi-impoverished noble family from northern France, he entered the royal army at age fifteen and soon developed an interest in the diverse local plants at his various postings in eastern France and along the Mediterranean coast. After leaving the army because of illness a decade later, Lamarck studied medicine and, thanks to Buffon, obtained minor positions at the French Academy of Science and the Royal Garden. When the Revolutionary government reorganized the Royal Garden into the Museum of Natural History in 1793, Lamarck landed the chair in invertebrate studies, which he held for the rest of his life. Regardless of his official duties, he mostly did what he wanted to do in science, which included publishing widely on all manner of topics in natural history. Plagued by poor health throughout his later life, Lamarck struggled with poverty, multiple children (many with physical or mental disabilities), and a succession of three or four wives (he married the first of them, who bore six of his children, on her deathbed and may have never married the last one). He became totally blind in 1818, and died destitute. Speculative in the extreme, Lamarck devised idiosyncratic views about science and religion that reflected the naive Enlightenment optimism of his youth. Of all his views, his transmutation hypothesis had the most lasting impact.

Although various aspects of Lamarck's thinking on transmutation evolved over time, the basic outline of his scheme remained much the same from the time that he devised it in 1800, at age fifty-five. Lamarck believed in the ongoing spontaneous generation of simple living organisms through action on physical matter by a material life force or fluid that he variously equated with electricity or nervous fluid in animals. This force or fluid could transform "gelatinous" matter into the simplest of animals, he claimed, and "gummy" matter into the simplest of plants. Once living organisms form (and this happens continuously, according to Lamarck), the fluid continues to act in them and their descendants—naturally driving them to evolve into ever more specialized forms. As seen by him but never widely accepted among scientists, the evolutionary process acts much like an ascending escalator in that the various types of organisms get on at different times but all ride it up simultaneously. This is progress, and Lamarck envisioned all living things participating in it.

According to Lamarck, the nervous fluid drives the evolution of animals in two basic ways. First, external stimuli and internal requirements can cause the fluid to concentrate in particular parts of the body through exercise, stimulating the emergence of a new organ there. Second, the fluid naturally flows toward used organs and away from unused ones, causing the former to develop further and the latter to atrophy. Lamarck saw the overall process as highly adaptive. By creating needs and stimulating organ use, long-term environmental change acting on a population would guide the development of organisms within that population in a particular direction. The best known example of Lamarckian evolution is that of the short-necked ancestors of the modern giraffe, which supposedly stretched their necks to reach ever higher tree leaves in response to drying conditions on the African savannah. As nervous fluid flowed to their necks, they grew longer. These changes would die with the individual unless they were passed on to their descendants. Thus, to make the process work, Lamarck posited that acquired characteristics fixed by the nervous fluid (such as a giraffe's longer neck) were heritable. As changes in individuals accumulated over generations, visibly different types evolved.

Aside from the fundamental difference between plants and animals, taxonomic distinctions (such as genus or species) lost any real meaning in a Lamarckian world. All organisms of every lineage were simply progressing toward greater complexity. This process might take differing forms in different organisms because of environmental conditions, of course, but the trend everywhere was the same. The current array of forms were neither fixed nor had common ancestry, Lamarck maintained, but were merely a snapshot of development over time from a multitude of beginnings, with more specialized organisms representing older linages than less specialized ones. The only constant was change.

In his lifetime, Cuvier had little difficulty marginalizing Lamarck and his transmutation hypothesis. The anatomic parts of living organisms are too interdependent—too finely designed—for the whole to evolve piecemeal, Cuvier explained, and the fossil record showed no sign of transitional forms. In 1830, Cuvier directed similar arrows against his colleague Geoffroy, who had the impertinence to revive evolutionary thinking at the Museum of Natural History in Paris. Of course, Cuvier never admitted to seeing progress in the fossil record, which made it easy for him to dismiss evolution. Still, when Grant and Chambers drew on emerging fossil evidence for progressive organic development to advance the idea of evolution, Sedgwick, Buckland, and other British catastrophists (who had uncovered much of that evidence) could effectively employ Cuvier's argument from design against it. Indeed, they sharpened it. For them, organic progress reflected the earth's directional geologic history. "Creation has been progressive on the whole—probably because [the Creator] adapted [it] to progressive conditions," Sedgwick wrote in 1850, "yet has this progress (when studied in detail) been so effected as to bid *defiance* to the pantheistic theory of development" espoused in Chambers's *Vestiges*.[13] For the British catastrophists, even more so than for Cuvier, each species so perfectly fit its time and place that it must have been designed by an intelligent Creator.

In their assaults against pre-Darwinian theories of evolution, Britain's Christian catastrophists had important allies. On the one hand, philosophical idealists such as zoologist Louis Agassiz and anatomist Richard Owen, both disciples of Cuvier, saw species as fixed ideas in the mind of the Creator. As such, they joined in opposing evolutionism. On the other hand, geologist Charles Lyell developed a theory of uniformitarianism that seemingly undermined the idea of progress in both geologic and biologic history. He led the charge against Chambers's *Vestiges*. Later, Darwin would draw on all of them to build a theory of evolution on a firmer foundation than that of his predecessors.

IDEALISM AND ARCHETYPES

Agassiz was a Swiss-born, German-educated ichthyologist who, by accepting a Harvard professorship in 1846, became the biggest fish in America's still small scientific pond. A charismatic public speaker and effective promoter of science, he reigned over American science for a quarter century. As a student in Germany, Agassiz imbibed the idealist vision, rooted in the philosophy of J. W. von Goethe, G.W.F. Hegel, and Lorenz Oken, that species constitute ideal, fixed forms related to one another in a coherent pattern of natural order. Biological idealists believed that the progressive appearance of species is recapitulated in embryonic development (such that the embryos of more complex or specialized organisms pass through the stages of simpler or more generalized ones) and displayed in the fossil record, with both these signposts pointing toward the ultimate emergence of the perfect human form (most likely a German, they presumed).

Agassiz's subsequent studies in Paris under Cuvier added a catastrophist framework to his idealist philosophy. After each catastrophe, Agassiz concluded, the Creator utilized ideal archetypes to recreate life according to a coherent, directional plan. "It is true that, taken as a whole, there is a gradation in the organized beings of successive geological formations, and that the end and aim of this development is the appearance of man," Agassiz affirmed in an 1845 letter supporting Sedgwick's stand against the early evolutionists. "But this serial connection of all successive creatures is not material; taken singly these groups of species *show no relation through intermediate forms* genetically derived one from the other. The connection between them becomes evident only when they are considered as a whole emanating from a creative power."[14] Certainly not an orthodox Christian and little interested in Sedgwick's notion that life progressed to fit improved geologic conditions, Agassiz could nevertheless agree, "The history of the earth proclaims its Creator."[15] Later, by championing this position even against Darwinism, Agassiz became something of a scientific saint to Christian creationists. For him, the progressive appearance of increasingly specialized species solely reflected their origin in the mind of God, not the impact of environmental factors or evolution.

Among the great British naturalists of the mid–nineteenth century (and there were many), none was more influenced by idealist thought than Richard Owen. A gifted comparative anatomist, by the late 1840s Owen came to see that certain basic forms underlay the structure of various vertebrate animals. He called them "homologies": cases where the same structure serves varying functions in different species. A human's hand, a bat's wing, and a whale's paddle share a common five-finger bone structure, for example, even though they perform different functions. In one sense, Owen's work undermined the creationist argument for intelligent design because no common structure can ever be optimal for multiple tasks. Indeed, Darwin would later use Owen's homologies as evidence for evolutionary adaption in related species and against creationism. For an idealist like Owen, however, homologies simply suggested that a single rational Creator employed one archetype to mold multiple species. This conception led Owen to see a branching pattern of directional development in species outward from a common archetype, rather than a straight line of progress from older to new species. Darwin integrated such a developmental pattern into his theory of evolution, which distinguished it from Lamarckian linearism. Owen thus helped shape modern ideas of directional organic development even though he originally rejected the very notion.

Like Cuvier, Owen began his scientific career as a comparative anatomist studying living specimens and only later moved into paleontology. Also like Cuvier, whom Owen saw as his model and mentor, he climbed from modest beginnings to the pinnacle of his profession, leaving many admirers but few friends along the way. Trained as a surgeon during the 1820s, Owen secured a position dissecting specimens for London's Royal College of Surgeons and gradually rose in its ranks by dint of his hard work and marrying the boss's daughter to become head of the college's natural-history museum. In 1856, he took command of the British Museum's natural-history collections, which he expanded into the world's finest. For his services to British science, Queen Victoria knighted him in 1884, and he died eight years later in lodgings provided by her.

Clearly troubled by the connection between belief in organic evolution and social radicalism, from the outset Owen (again like Cuvier) turned his science against the transmutation hypothesis, going so far during the 1830s and early 1840s as to deny seeing any directional or progressive trend in the fossil record. British transmutationist Robert Grant became Owen's bête noire and, beginning in the 1830s, Owen savaged him as viciously as Cuvier did Geoffroy. Owen made a particular point of positing a complete anatomic break between humans and apes, a finding that would later draw fire from evolutionist Thomas Henry Huxley. For all orders of animals, Owen tried to show that earlier species were not necessarily less complex or specialized than later ones. He did this most spectacularly in his 1841 "Report on British Fossil Reptiles," in which he transformed the giant lizards first identified by Buckland and Mantell into a new order of beast, which he gave the suitably arresting name *Dinosauria,* or "terrible lizard."

Buckland and Mantell were content to leave their giant lizards crawling on the ground like the familiar, cold-blooded reptiles of today. Owen raised them from the ground on mammal-like legs and claimed that each had "a highly-organized centre of circulation in a degree more nearly approaching that which now characterizes the warm-blooded Vertebrata."[16] They became the pinnacle of reptilian life, far superior to any earlier or later lizard and a dramatic interruption in anyone's concept of progressive development. With this and other interpretations of fossil reptiles that he provided in his 1841 report (interpretations that "a too cautious observer would, perhaps, have shrunk from," he observed), Owen concluded that "though a general progression may be discerned, the interruptions [of it] negative the notion that the progression has been the result of self-developing energies adequate to a transmutation of specific characters; but, on the contrary, support the conclusion that the modifications of [anatomic] structure which characterizes the extinct Reptiles were originally impressed upon them at their creation."[17] So much for the transmutation hypothesis: According to Owen, God specially created dinosaurs.

Owen imposed similarly antiprogressivist interpretations on other extinct species until the late 1840s, when he finally began detecting a branching pattern of directional development in the fossil record. Still, his idealist orientation allowed him to see a nonevolutionary cause for this pattern. Throughout, he ambitiously used his scientific brilliance and museum position to appropriate the findings of other naturalists and give them his own interpretation. "It is astonishing with what an intense feeling of hatred Owen is regarded by the majority of his contemporaries," Huxley observed in 1851, "with Mantell as arch-hater."[18] Owen thought of himself as England's Cuvier—and in too many ways, he was.

THE UNIFORMITARIAN CHALLENGE

For a while, Charles Lyell joined Owen in using evidence of apparent regression in the fossil record to counter evolutionism. Indeed, second only to Owen, Lyell was the most prominent British naturalist to adopt such a tactic. As with Owen's biologic homologies, however, Lyell's geologic uniformitarianism ultimately helped make the case for Darwinism.

Called to the bar in 1822, Lyell grew quickly bored of practicing law and opted to make his name in geology—a subject that had fascinated him since he attended two of Buckland's courses as an Oxford student in 1817 and 1818. As ambitious as Owen, though not as obnoxiously so, Lyell consciously (and with apparent sincerity) chose to champion a geologic theory that was at once revolutionary enough to excite interest yet subversive of the socially radical transmutation hypothesis. He found the seeds of that theory in the "steady-state vulcanism" of the late-eighteenth-century Scottish gentleman-scientist James Hutton, which Lyell updated with fossil and other evidence to create the modern theory of uniformitarian geology. If he could make the case for uniformitarianism, Lyell reasoned, and as a barrister he thought that he could, then his fame would be secure. Lyell fully appreciated the challenge before him, and confidently accepted it. "The fact is that to become great in science a man must be nearly as devoted as a lawyer, and must have more than mere talent," he warned his fiancée in 1831.[19] Lyell picked the path of scientific greatness, and it led him to a knighthood.

Before Lyell, mainstream scientific opinion gave a directional orientation to geologic history. The earth was hotter and wetter in the distant past, most naturalists believed, and past geologic events were more dramatic than present ones. The influential late-eighteenth-century German mineralogist Abraham Werner and his so-called "Neptunist school" saw rock strata, fossils, and geologic features formed through the gradual retreat of a vast, primordial ocean. To better account for discontinuities in the fossil record, Cuvier and his followers believed that the oceans came and went (or at least shifted around) in earlier epochs, probably with decreasing intensity. Other late-eighteenth- or early-nineteenth-century naturalists speculated about massive prehistoric earthquakes, volcanos, or glaciers shaping the earth's features. For all these naturalists, from the devoutest Christians to the rankest materialists, geologic history had a beginning, a direction, and a probable end. Living for the most part in the now-geologically quiet regions of northern and western Europe, they could not conceive of current geologic forces shaping the earth's features and causing discontinuities in the fossil record. Those forces must have been greater in the past.

On philosophical grounds, James Hutton could not accept such a world. A deist, he believed that God should have cre-

ated a self-sustaining earth—an ideal home for humans—
that suffered no permanent directional change. A radical empiricist or "actualist," he sought to explain currently
observable geologic features through the operation of currently observable geologic forces. In his concept of steady-
state vulcanism, which satisfied both these requirements,
Hutton proposed a cyclical process of igneous-rock mountains gradually rising from the earth's molten core and then
slowly weathering to create inhabitable land. As this land accumulated over time, its bottom layers would push down toward the fiery core and remelt. The pressure resulting from
accumulating layers would push up new mountains from the
molten core, and the deliberate, law-bound cycle would repeat itself. "It was this train of reasoning," Lyell later explained, "which induced the Scottish geologist Hutton to
declare that in the economy of the world 'he could find no
trace of a beginning, no prospect of an end.' "[20]

Schooled in Buckland's catastrophism, Lyell embraced
steady-state vulcanism with a convert's zeal beginning in the
late 1820s, and transformed it into modern uniformitarianism.
He had at least two motivations. First, he sincerely accepted
the view that science should only employ known naturalistic
causes operating in observable ways to explain natural phenomena. Invoking larger-than-life past catastrophes smacked
of religion, Lyell argued. Second, he believed that a nondi-
rectional geologic history would undermine Lamarckism,
which he saw as dangerously subversive of human dignity.[21]
In their moral superiority and reasoning powers, humans
stood too far above beasts to have ever descended from them,
Lyell maintained.

In his three-volume *Principles of Geology*, first published
from 1830 to 1833, Lyell refashioned Hutton's cyclic outline
of geologic history into a coherent scientific theory. Using
the polemic skills of a barrister to make his case, Lyell wove

in observations from Italy and other volcanically active or earthquake-prone regions to suggest that (given limitless time) the earth's inner heat could dramatically sculpt geologic features. He also turned the fossil record to his advantage by stressing that the breaks in it were neither so complete nor so dramatic as catastrophists claimed. Further, while conceding that individual species appeared and disappeared over time, Lyell denied that the overall pattern was as progressive as assumed by most British catastrophists and required under Lamarckism. Quite to the contrary, he argued (on the basis of a few disputed examples) that representatives from all classes of plants and animals appeared throughout the fossil record. Indeed, although he agreed with catastrophists that God specially created species to fit their environments, he saw long-term environmental change (and therefore historical changes in the flora and fauna) as gradual and cyclical rather than abrupt and directional. An age dominated by mammals could just as easily precede as follow one dominated by reptiles, Lyell believed; it simply depended on environmental conditions at the time. Of course the fossil record did not fully display such a cyclical pattern of life, but Lyell attributed this to its incompleteness. Fossils are only laid down in particular conditions, he noted, and are eventually destroyed in the cyclical subsidence of older rock strata. Throughout his pre-1860 writings and lectures, Lyell stressed the unique place of humans in creation and denied the possibility of evolution.

Through his books and lectures, Lyell reached a broad popular audience with his arguments, but he had mixed results persuading other naturalists. From reading Lyell's work, some of them became more hesitant to invoke the supernatural to account for the natural, and more willing to acknowledge the role of currently observable geologic forces in shaping the earth's features. Among geologists, catastrophism ultimately lost favor. For Darwin, uniformitarianism greatly

lengthened the time available for evolution to operate and il-
lustrated the cumulative power of small changes. In this
sense, as Huxley later observed, Lyell was "doomed to help
the cause he hated."[22] But Lyell made little headway in eliminating the profession's or the public's perception of progress
in the fossil record. Murchison's publications repeatedly took
Lyell to task for denying direction in the fossil record, for example, and Sedgwick wrote of him, "He is an excellent and
thoughtful writer, but not, I think, a great field-observer, and
during his long geological labours he has never been able to
look steadily in the face of nature except through the specta-
cles of an hypothesis."[23] Lyell became especially bitter in
1851, when even Owen publically turned against him on the
matter of progress in the fossil record, yet Lyell held his
ground for another decade, until he guardedly accepted a
Darwinian version of transmutation.

———

By the 1850s, the issue of moment in biology was evolution.
Among naturalists, political conservatives (such as Cuvier
and Owen) and religious traditionalists (such as Buckland and
Sedgwick) instinctively opposed the idea that species evolve,
at least in part because it normalized change. Largely for the
same reason, social radicals (like Lamarck and Chambers) inevitably embraced it. Partisans on either side of this great di-
vide could muster enough scientific evidence to make their
positions plausible.

   Although Cuvier's catastrophist interpretation of the fossil
record stood as a bulwark for creationism in the life sciences
for a generation, it sowed the seeds of its own destruction. On
the one hand, many naturalists inspired by Cuvier's basic outline of the earth's history, including Buckland, Sedgwick, and
Murchison, began sensing directional change in the life forms
of succeeding epochs. For naturalists outside Cuvier's fold,
most notably Lamarck and Geoffroy, the apparent progress in

biologic history was all but overwhelming. On the other hand, scientists inclined toward methodological naturalism, personified in Britain by Charles Lyell, questioned Cuvier's invocation of unobserved past catastrophes to account for the pattern of species in the fossil record. They saw Cuvier use empirical evidence from nature to induce the existence of past geologic epochs, then call upon unobserved forces to deduce the connections among them. He might as well have invoked miracles, they complained, which would be positively unscientific.

Methodological naturalism limited scientists to the task of looking for natural (as opposed to supernatural or other non-natural) causes for physical phenomena, and left the rest to theologians, philosophers, historians, or other scholars. Indeed, the very term "scientist" was coined during the 1830s in part to distinguish persons engaged in this pursuit from other scholars, replacing such earlier designations as "natural philosophers" (for those studying physics) and "natural historians" (for those studying the life sciences). Once fully integrated, an appreciation of organic succession coupled with a sense of direction in biologic history and a commitment to methodological naturalism in science made the acceptance of evolution in biology virtually inevitable. The discovery and identification of dinosaurs as an earlier order of dominant land animal simply hastened and popularized the emerging scientific view of organic succession and development over time. It was time for Darwin to embark on his voyage of scientific discovery.

CHAPTER 3

ON THE ORIGINS
OF DARWINISM

After repeated delays due to heavy gales in the Channel, the diminutive British Navy vessel H.M.S. *Beagle* set sail from Plymouth, England, on December 27, 1831, for a projected two-year expedition to chart the southern coast of South America and, perhaps, the islands of the South Pacific. For Charles Darwin, the expedition's young naturalist, it proved an inauspicious start for what became a five-year voyage that would shape his professional career and thrust him into the center of the storm over the origin of species.

The seas ran high for the first week, and harsh punishments inflicted on the crew for predeparture holiday drunkenness made those initial days almost hellish to the well-bred naturalist. "Waked in the morning with an eight knot per hour wind, & soon became sick & remained so during the whole day—my thoughts most unpleasantly occupied with the flogging of several men for offenses brought on by the indulgence granted them on Christmas day," Darwin wrote in his diary about his first full day at sea. For the second day, he added, "The misery is excessive & far exceeds what a person would suppose who had never been at sea more than a few days." The ship's aristocratic captain, Robert FitzRoy, feared that Darwin would abandon the adventure at first landfall. The thought crossed Darwin's mind, as well. "I often said before starting, that I had no doubt I should frequently repent of the whole undertaking, little did I think with what fervour I should do so," Darwin wrote on day three. "I can scarcely conceive any more miserable state, than when such dark & gloomy thoughts are haunting the mind as have to day pursued me."[1]

The choice of Darwin for the expedition and his willingness to go reflected the scientific culture of nineteenth-century Britain. Government-sponsored voyages of scientific discovery had become commonplace by this time. Captains James Cook and George Vancouver had circumnavigated the globe in tall ships with teams of scientists charting the coasts, making scientific observations and collecting natural-history specimens for Britain during the late eighteenth century. France and other European powers had countered with similar expeditions of their own. Even the new United States government was preparing to launch such an endeavor later in the 1830s. Formal graduate programs did not yet exist; in their absence, many of the nineteenth century's finest naturalists cut their teeth aboard scientific expeditions—before settling into permanent positions at universities, natural-history museums, and other institutions.

The voyage of the *Beagle* was not planned as a grand expedition for science, though it later became one. Indeed, it did not even merit an official naturalist. A ninety-foot-long brig awkwardly fitted with three masts, the *Beagle* was better suited to poke along a coast than to sail the high seas. In 1830, it returned prematurely from a mission to chart the coasts in and around the southern tip of South America after its melancholic captain, lonely in command and lost in the desolate labyrinth of Tierra del Fuego, killed himself. The British admiral in charge of the South American station sent FitzRoy, not yet twenty-five but a direct descendant of King Charles II, to assume command of the brig and, ultimately, return it to Britain. Two years later, with FitzRoy still in command, the *Beagle* headed back to finish its mission, with authority to continue on around the world. The young commander suffered a similar temperament as his predecessor, however, and feared a similar fate. FitzRoy's uncle, Lord Castlereagh, had slit his own throat, and FitzRoy would do the same in 1865. For the

Beagle's voyage, he wanted someone on board he could talk with as an equal—and thus he secured permission to take along a gentleman naturalist.

Although he was no one's first choice for the position, Darwin filled its peculiar requirements. Born in 1809 into an affluent family of English capitalists living in rural Shropshire, Darwin was five years younger than FitzRoy and had just graduated from Cambridge. He had developed an abiding interest in natural history at Cambridge, where he regularly collected plants with the young botany professor John Stevens Henslow and once accompanied Adam Sedgwick on a geology field trip. Graduation left Darwin with only vague plans for his future, and the prospect of a round-the-world scientific expedition, even one on the *Beagle*, interested him greatly. Darwin's father initially resisted—seeing the voyage as simply another expensive dalliance for his capricious son—but soon relented (as he usually did) and underwrote the entire cost for his son and (ultimately) his son's manservant, Syms Covington.

"Gloria in excelsis is the most moderate beginning I can think of," Darwin wrote to Henslow, who had recommended Darwin for the position. "What changes I have had: till one to day I was building castles in the air about hunting Foxes in Shropshire, now Lamas in South America. There is indeed a tide in the affairs of men."[2] Visions of the South Pacific especially excited him. "It is such capital fun ordering things," Darwin wrote four days later to a college chum, "to day I ordered a Rifle & 2 pair of pistols; for we shall have plenty of fighting with those d—— Cannibals: It would be something to shoot the King of the Cannibals Islands."[3] Delays and the first week's seasickness dimmed his early enthusiasm, but by the time the *Beagle* reached its first planned landfall, the Canary Islands southwest of Spain, Darwin had recovered his characteristic exuberance. "We saw the sun rise behind the

Charles Darwin, from a portrait made in 1840, shortly after his return from the voyage of the *Beagle*.

rugged outline of the Grand Canary island, and suddenly illumine the Peak of Teneriffe," he wrote. "This was the first of many delightful days never to be forgotten."[4]

A cholera quarantine kept the expedition from disembarking on the Canary Islands, so the *Beagle* sailed farther south to the tropical Cape Verde Islands instead. The expedition's layover in that volcanic archipelago transformed Darwin's thinking about geology. He had begun the voyage as a conventional British catastrophist still under Sedgwick's sway. Indeed, Sedgwick had supplied Darwin with a reading list for the voyage—a list that conspicuously omitted Lyell's controversial *Principles of Geology*. FitzRoy gave a copy of the book to

Darwin, however, and the young naturalist was reading it when the expedition landed on St. Jago, the largest island in the Cape Verde group. What Darwin saw there made him an instant and lifelong convert to Lyellian uniformitarianism.

"On entering the harbour, a perfectly horizontal white band in the face of the sea cliff, may be seen running for some miles along the coast, and at the height of about forty-five feet above the water," Darwin wrote in his *Journal*. Upon close examination, he found that the formation consisted of a light layer of rock derived from cooked corals and seashells between dark layers of volcanic rock. The sea life that created the white band must have lived on a shallow shoal of volcanic rock and been covered by a flow of molten lava while still submerged, Darwin surmised. Then the entire formation rose to its present height gradually enough to maintain its shape. All this must have happened long ago, because the island's volcanic craters were weathered almost beyond recognition; but not *too* long ago, because the shells in the white band were of the same types as on the beach below. Within his first few days on St. Jago, the twenty-two-year-old naturalist had interpreted to his own satisfaction the geologic history of the Cape Verde Islands using Lyellian activism rather than Cuvierian catastrophism. Current geologic forces operating over time could have produced these islands, Darwin concluded, whereas catastrophic past events would have disrupted the strata.[5]

Deciphering island geology was a heady and empowering experience for Darwin. Upon disembarking on St. Jago, he was so "overwhelmed" by the island's unfamiliar volcanic terrain and tropical plants that he wrote in his diary, "It has been for me a glorious day, like giving to a blind man eyes."[6] Recalling the episode nearly a half century later, Darwin wrote that the cliffs of St. Jago "showed me clearly the wonderful superiority of Lyell's manner of treating geology, compared to that of any other author." Suddenly, his vision of the voyage's scientific significance and of himself as a scientist enlarged. "It then first dawned on me that I might perhaps write a book on the geology of the various countries visited, and this made me thrill with delight," Darwin related. "That was a memorable hour for me, and how distinctly I can call to mind the low cliff of lava beneath which I rested, with the sun glaring hot, a few strange desert plants growing near, and with living corals in the tidal pools at my feet."[7] At that moment, the student-traveler became a self-confident scientist.

Subsequent observations reconfirmed Darwin's newfound faith in geologic actualism. From interpreting coral reefs as the product of gradual subsidence to witnessing active volcanos forming the Galápagos Islands, Darwin saw evidence all around him of the profound effect of ongoing natural forces. Catastrophists could account for these observations by invoking prehistoric events beyond the magnitude of current ones, of course, but such explanations no longer satisfied Darwin—particularly after he experienced a major earthquake in Chile. "The motion made me giddy," he noted in his diary for February 20, 1835. "The world, the very emblem of all that is solid, moves beneath our feet like a crust over a fluid."[8] For Darwin, the earthquake proved the mountain-building power of current geologic forces. "Captain Fitz Roy found beds of putrid mussel-shells *still adhering to the rocks,* ten feet above high-water mark," Darwin observed with emphasis in his *Journal* about one site the expedition visited two weeks after the earthquake. "The elevation of this province is particularly interesting, from its having been the theatre of several other violent earthquakes, and from the vast numbers of sea-shells scattered over the land, up to a height of certainly 600, and, I believe, of 1000 feet."[9] About such an earthquake, Darwin wrote to a friend shortly after experiencing the Chilean one,

"It is certainly one of the very grandest phenomena to which the globe is subject."[10]

———

Although Darwin's conversion to uniformitarian geology lay the foundation for his later acceptance of organic evolution, the second step did not follow automatically from the first. Indeed, Lyell himself long maintained that uniformitarianism (by denying direction in geologic history) affirmatively undermined evolutionism. In *Principles of Geology*, he offered the alternative gradualist view that God (or a "Presiding Mind") continually created species to fit local environments. According to this view, those species would spread out from their "*centre* or *foci* of creation" to occupy suitable territory for so long as environmental conditions permitted, and then become extinct.[11] Darwin spent much of his time during the *Beagle* expedition looking for the Lyellian "centre of creation" for individual species, and interpreting the distribution of various plants and animals accordingly.[12] Yet by lengthening the earth's history indefinitely, eliminating life-destroying catastrophes, and postulating gradual environmental change over time (which presumed gradual organic change, as well), a uniformitarian view of geology points those committed to its principles toward an evolutionary view of biology. Darwin the disciple simply surpassed Lyell the master in accepting the implications of uniformitarianism.

Absorbing Lyell's *Principles of Geology* during the *Beagle* expedition affected Darwin in subtle ways, as well—so much so that he dedicated his book about the voyage to Lyell, "as an acknowledgment that the chief part of whatever scientific merit this *Journal* and the other works of the author may possess, has been derived from studying the well-known and admirable *Principles of Geology*."[13] For instance, Lyell opened the *Principles* with a self-serving history of geology that uncritically hailed any insight anticipating uniformitarianism and utterly dismissed the contributions of catastrophists. This account heaped particular scorn on revealed religion and church doctrine for holding back scientific progress. "In short," Lyell concluded near its end, "a sketch of the progress of geology is the history of a constant and violent struggle between new opinion and ancient doctrines, sanctioned by the implicit faith of many generations, and supposed to rest on scriptural authority."[14] This was law-office history written by a barrister, and Darwin swallowed it whole. In its methods and findings, however, the geology of Lyell did not represent a revolutionary advance over that of Sedgwick, Murchison, Agassiz, or Owen (all of whom favored naturalistic explanations for geologic phenomena and helped lay the foundation for the modern understanding of the geologic column), but the *Principles* made it seem so—and Darwin signed on for the revolution. "I always feel as if my books came half out of Lyell's brains," Darwin later wrote, "for I have always thought that the great merit of the *Principles*, was that it altered the whole tone of one's mind & therefore that when seeing a thing never seen by Lyell, one yet saw it partially through his eyes."[15]

———

Just as the observations that Darwin made on the Cape Verde Islands early in the *Beagle* expedition opened his eyes to uniformitarian geology, what he saw in the Galápagos archipelago late in the voyage inspired his thoughts on organic evolution. Like the Cape Verde group, the Galápagos Islands are isolated and desolate. Seeing them through Lyell's eyes, Darwin recognized both archipelagos as the peaks of volcanic mountains risen relatively recently from the sea. They remained hostile living environments with a limited variety of mostly indigenous species. The Cape Verde Islands stand at fifteen degrees north latitude nearly four hundred miles off the Atlantic coast of Africa; the Galápagos straddle the equator more than five hundred miles off the Pacific coast of South America. The physical ecology of the two places was similar, yet the former's plant and animal species were like those of Africa and the latter's like those of South America. "The creationist" must consider these "as so many ultimate facts," Darwin wrote in a private 1844 essay that summarized his thinking of the previous eight years. "He can only say that it so pleased the Creator... that the inhabitants of the Galapagos Archipelago should be related to those of Chile... and that all its inhabitants should be totally unlike those of the similarly volcanic and arid Cape de Verde and Canary Islands." This could be, Darwin conceded, "but it is absolutely opposed to every analogy, drawn from [physics] that facts, when connected, should be considered as ultimate and not the direct consequence of more general laws." In short, he charged, the creationist explanation was unscientific.[16]

In 1837, Darwin began outlining his evolutionary explanation for these observations in a series of private notebooks and essays, one of which identified the "species of the Galapagos Archipelago" as a primary source "of all my views."[17] In his first such notebook, for example, he jotted, "My idea of Volcanic islands elevated. Then peculiar plants created.... Yet new creation affected by Halo of neighboring continent."[18] In other words, individual plants and animals from the nearest landmass should colonize a newly formed island, become isolated there from their parent population, and then evolve to fit the island's environment and fill available niches. Spelling this out in an 1842 essay, Darwin noted, "So if Island formed near continent, let it be ever so different that continent would supply inhabitants, and new species (like the old) would be allied with that continent."[19] Addressing the same issue in an 1838 notebook entry, he asked himself rhetorically, "Did Creator make all new [species on oceanic islands,] yet

[with] forms like [on] neighbouring Continent? This fact speaks volumes. My theory explains this but no other will."[20] Only through a process of colonization, isolation, and evolution would the Cape Verde Islands have African-like species and the Galápagos Islands have American-like species, Darwin reasoned. A Creator would have fashioned species to fit their environment, not some neighboring continental template.

While such reasoning reinforced Darwin's thinking about evolution, his initial conversion experience came from the even narrower observation that these relationships between South American and Galápagos species carried over to inter-island differences. Although he did not notice it while collecting specimens in the archipelago from September 16 to October 20, 1835, on closely examining them during the *Beagle*'s ensuing yearlong voyage back to Britain, Darwin recognized a potentially significant relationship among Galápagos mockingbirds. "I have specimens from four of the larger Islands," he recorded in a notebook he kept aboard ship. "The specimens from Chatham and Albemarle Islands appear to be the same, but the other two are different. In each island each kind is *exclusively* found." If these specimens represented distinct species (as opposed to simply marked varieties of the same species), then those species must have evolved in isolation on their separate islands from a common ancestral type (probably blown to the archipelago from South America), Darwin surmised. "Such facts would undermine the stability of species," he concluded.[21]

Soon after the *Beagle* returned to Britain late in 1836, ornithologist John Gould definitively identified three island-specific species of Galápagos mockingbirds from Darwin's specimens. Even more striking, he identified fourteen species of Galápagos finches (differentiated primarily by the size and shape of their beaks) from the array of small land birds in

Charles Darwin's 1845 sketch of Galápagos finch beaks.

Darwin's collection. Darwin could not determine from his records whether the various finch species came from separate islands, but if some of them did (and he believed it likely), then they reinforced his conclusion. A rational Creator would not have made so many different species of finches and mockingbirds for ecologically similar islands in one small archipelago, he argued to himself. Adaption to niches became Darwin's answer for the origin of these species and, ultimately, by extrapolation, of all species everywhere.

By themselves, however, isolation and evolution represented only a partial answer to what Darwin called in his travel *Journal* "that mystery of mysteries—the first appearance of new beings on this earth."[22] They did not explain how the process operated. Doing so became Darwin's obsession for the rest of his life. He knew that other naturalists had proposed theories of organic evolution before, but none of them contained a credible mechanism for evolutionary change. For forty years, mainstream scientists had ridiculed Lamarck's theory that living things evolve by adapting to their changing environment and passing on those adaptations to their descendants. Similar theories offered by others—including a pre-Lamarckian one by Charles Darwin's grandfather, the poet, philosopher, and physician Erasmus Darwin—suffered a similar fate, or were simply ignored by scientists. Lamarck, Erasmus Darwin, and many of the other early evolutionists seemed to revel in their outcast status—but Charles Darwin had no stomach for it. He craved acceptance within the scientific community even as he sought to overthrow one of its most basic beliefs: the fixity of species. So he labored on his theory virtually in secret for two decades, all the while keeping up appearances as a conventional Victorian gentleman scientist.

Darwin played the part well. In 1839, he married his wealthy first cousin, Emma Wedgwood, whose dowry (when coupled with his own family fortune) eliminated any need for him to earn a living. Three years later, the couple moved to a country home in Downe, which was near enough to London for him to participate (when he wanted) in its scientific culture, but rural enough for him to conduct breeding experiments with domesticated plants and animals designed to study the evolutionary process. Beginning soon after his return to Britain in 1836 and continuing until near his death in 1882, he produced a steady stream of scientific books and articles. Addressing a wide variety of topics in geology, biology, and psychology—from barnacles and South American fossils to pigeon breeding and the expression of emotions—they all contributed to Darwin's understanding of evolution. His scholarly publications and social standing gained him entry into Britain's elite institutions of science, including election at age thirty to the prestigious Royal Society of London.

———

Darwin's conceptual breakthrough came in 1838, after he began considering the case of human evolution. For most people concerned about the issue of transmutation (either pro or con), the key question is always the same: Did humans evolve from other primates? Darwin knew the literature on this subject, of course, and had directly confronted the matter during the *Beagle* expedition when he encountered the native peoples of Tierra del Fuego, who he deemed the lowest form of humanity on earth.[23] In 1838, while struggling to understand how evolution worked, Darwin's thoughts returned to the Fuegians and their apparent similarity to primates in the London zoo.

Those and other thoughts exploring supposed links between humans and animals pepper his private notebooks throughout 1838. "Let man visit orangutan in domestication, hear expressive whine, see its intelligence," Darwin wrote early in the year, "then let him dare to boast of his proud preeminence." Here he inserted the phrase, "not understanding language of Fuegian[s], puts [them] on par with Monkeys." Returning to this comparison later in the year, he added, "Forget the use of language, & judge only by what you see. Compare, the Fuegian & Orangutan, & dare to say difference so great." Darwin downplayed the language factor, as well. "The distinction as often said of language in man is very great from all animals—but do not overrate—animals communicate to each other," he noted in one of many entries attributing supposedly human powers to beasts. Just as frequently he speculated about animal origins of human traits, such as when he wrote, "One's tendency to kiss, & almost to bite, that which one sexually loves is probably . . . due to our distant ancestors have been like *dogs* to bitches." As for the vaunted "mind of man," Darwin concluded, it "is no more perfect, than instincts of animals." Human thought itself (like animal instincts) he attributed to brain structure, chiding himself "oh you Materialist!" for thinking so. Continually he probed the perceived boundary questions. "Does a negress blush? I am almost sure [the Fuegians] did," he asks himself at one point. "Animals I should think would not."[24]

Absorbed by such comparisons, Darwin immersed himself in books and articles about animal aspects of the human condition. During the course of this reading, he took up Thomas Malthus's classic *Essay on the Principle of Population*. All species, including humans, reproduce at unsustainably high rates, Malthus asserted. Lacking sufficient food to go around, "necessity, that imperious all pervading law of nature, restrains them within the prescribed bounds," he explained. "Among plants and animals its effects are waste of seed, sickness, and premature death. Among mankind, misery and vice."[25]

The practical implications of Malthus's so-called "principle of population" are profound, complex, and controversial. Focusing on humanity in his *Essay*, Malthus used it to argue against welfare programs for the poor, presenting handouts as a recipe for added human suffering in the long run. Extending the principle to all living things, Darwin saw in it a natural mechanism for evolutionary development. Beginning with the assumption that all individuals of every species naturally differ, he surmised that within each species a competitive struggle for existence would eliminate the weaker members and leave the stronger (or better adapted) ones to reproduce and pass along their beneficial adaptations to the next generation. "One may say there is a force like a hundred thousand wedges trying [to] force every kind of adaptive structure into gaps in the economy of Nature, or rather forming gaps by thrusting out weaker ones," Darwin wrote in a notebook entry dated September 28, 1838. "The final cause of all this wedging, must be to sort out proper structure & adapt it to change."[26] Describing the rush of comprehension four decades later in his *Autobiography*, Darwin remembered suddenly realizing that he "had at last got a theory by which to work."[27] He called his theory "natural selection."

Darwin equated the process to the artificial selection methods utilized by plant and animal breeders. These breeders created and sustained highly differentiated varieties by continually selecting for certain desired traits in the breeding stock, such as long ears in basset hounds or creamy milk in Jersey cows. Reasoning by analogy, Darwin saw intraspecies competition for food and mates creating new species within a given environment by continually selecting for traits that contributed to survival and reproduction, such as large, strong beaks for birds in places with hard seeds. "It is a beautiful part of my theory," he noted in late 1838 or early 1839, "that domesticated races of organisms are made by precisely [the] same means as species—but [the] latter far more perfectly & infinitely slower."[28]

For Darwin, a species simply constituted a population of physically similar individuals capable of breeding together, not an ideal, unchanging life-form. These similar (but not identical) individuals would necessarily compete with one another for the same limited resources in a Malthusian world, leaving the fittest among them to survive and reproduce their kind. He realized that different natural environments and ecological niches would favor different adaptations, so that species would not evolve in a linear, Lamarckian fashion. Rather, Darwin envisioned a branching process of evolutionary development, with various daughter species evolving in different directions from a common ancestral type to fill available geographic spaces and ecological niches. For Darwin, differential death rates caused by purely natural factors created new species. God was superfluous to the process.

Indeed, God became more than superfluous under Darwin's emerging view of origins—He became problematic. At the very least, the theory of evolution dispenses with the immediate need for a Creator to shape individual species, including humans. More critically, a natural-selection mechanism relying on cutthroat intraspecies competition to evolve new species struck Darwin as incompatible with any reasonable notion of benevolent divine action. Darwin's long drift toward agnosticism gained momentum at this point in his intellectual pilgrimage, and perhaps was accelerated later by such personal experiences as his worsening physical ailments and the death in 1851 of his beloved ten-year-old daughter, Annie.[29] In his private notes, Darwin began attributing religious belief to instinct and love of God to brain organization. As for humans, he wrote shortly after his Malthusian breakthrough, "When two races of men meet, they act precisely like two species of animals—they fight, eat each other, bring diseases to each other, and etcetera, but then comes the more deadly struggle, namely which have the best fitted organization or instincts (i.e. intellect in man) to gain the day."[30] In his mind's eye, Darwin surely saw the forces of British imperialism triumphing across the globe.

Essential to Darwin's conception was a modern worldview influenced by ideas of utilitarianism, individualism, imperialism, and *laissez-faire* capitalism. Of course Malthus was a utilitarian-minded political economist who championed the *laissez-faire* ideal. Darwin also read the writings of Adam Smith and other utilitarian economists who presented individual competition as the driving force of economic progress. Perhaps more important, he lived in a society that embraced this view; Darwin himself came from a family of successful capitalists. Further, he rode on the rising tide of British economic, political, and cultural imperialism when he sailed aboard the *Beagle*. "In the unknown interlocking movements of the human mind," Darwin biographer Janet Browne concludes, "natural selection intuitively seemed the right answer to a man thoroughly immersed in the productive, competitive world of early Victorian England."[31]

Darwin conceived his theory in 1838, but he did not publish anything about it for twenty years. Recognizing the depth of opposition among scientists to the transmutation hypothesis, he spent much of this time endeavoring to anticipate and answer in advance objections to his theory. In the process, he perfected his thinking on the gradual divergence of varieties into distinct species through competition, marshaled evidence for evolution from comparative anatomy and embryology, fitted fossils into evolutionary series and distribution patterns, and tried to imagine intermediate stages in the development of the eye and other complex organs. Although largely irreligious himself, he also worried about the impact that revealing his theory might have on religious believers, particularly his wife. "What a book a Devil's chaplain might write on the clumsy, wasteful, blundering low & horridly cruel works of nature!" Darwin exclaimed in an 1856 letter to British botanist Joseph Hooker, as if to justify not doing so.[32] Hooker was one of the few scientists that Darwin told about his theory prior to announcing it publicly. Lyell and American botanist Asa Gray, who supplied Darwin with information about the geographic distributions of Pacific-Rim plant species, were two others.[33] All three of these key confidants expressed interest in Darwin's theory, and Lyell urged him to publish it promptly in full, but none of them was yet ready to abandon creationism.

———

By the 1850s, however, British public opinion was warming to the idea of evolution. Despite hostile reviews, Robert Chambers's 1844 *Vestiges of the Natural History of Creation* sold enormously well for more than a decade and stimulated widespread discussion of human evolution. Beginning with his 1851 book, *Social Statics,* the popular British philosopher Herbert Spencer picked up where *Vestiges* left off in linking an essentially Lamarckian view of organic evolution with a Malthusian vision of human social progress through struggle

Alfred Russel Wallace,
from a daguerreotype
made in 1848, shortly
before his departure
for the Amazon basin.

and competition. It was Spencer, not Darwin, who coined the
term "survival of the fittest." Then, on June 18, 1858, Darwin
received a manuscript from evolutionist Alfred Russel Wallace containing the core concepts of natural selection. Darwin would have to publish his theory or risk losing priority.

Like Darwin, Wallace was fairly well-known among British
naturalists even before the joint announcement of their grand
theory in 1858. Although the two men differed in background
and temperament, they hit upon the idea of natural selection
in nearly the same way. The parallels and perpendicularities
between them are striking. Wallace grew up poor in rural
Britain and was largely self-educated, whereas Darwin received the best education that money could buy. In their late
teens, both men became fascinated with natural history. Exploiting the means open to them, both transformed this hobby
into a career. Capitalizing on his social standing, family
wealth, and university training, Darwin was chosen for the

Beagle expedition and thereafter settled comfortably into the life of a gentleman scientist. He never had to earn a living. From 1848 to 1862, Wallace embarked on bare-bones collecting trips, first to the Amazon basin with fellow collector Henry Walter Bates, and then to the Malay Archipelago, paying his own way by shipping back animal skins, pressed insects, and dried plants for sale to British collectors. He traveled by commercial steam or sailing ship from port to port and with native guides to places where few Europeans had ever gone. The inverse of Darwin, Wallace instinctively felt an affinity for the native peoples he encountered and a certain distance from European colonialists. Both men first gained scientific recognition for the specimens they sent or brought to Britain, and then secured wider fame by publishing popular accounts of their travels.

The transmutation hypothesis was widely debated but little accepted among European naturalists during the early nineteenth century. It had a revolutionary taint. Predisposed to embrace radical ideas in politics, religion, and science, Wallace instinctively accepted the idea that everything evolved; he pushed the range of his collecting trips in part to test his hypothesis that, under an evolutionary distribution pattern, similar (or nearly related) species should inhabit neighboring territories. This, he hoped, would serve as persuasive evidence for organic evolution. More conservative in his thinking, Darwin stumbled on just such a pattern for similar species of birds on the Galápagos Islands in 1835, and privately concluded that they must have evolved from a common ancestral type. Once convinced that species evolve, he was even more dogged than Wallace in trying to understand the process.

Their shared interest in the geographical distribution of species led Darwin and Wallace to begin corresponding in 1855, but neither initially told the other about his obsession with finding the mechanism driving evolution. Each found his answer in Malthus. It came to Darwin in a comfortable London salon; it struck Wallace during the height of a malarial fever in a native hut at the village of Dodinga on what is now the Indonesian island of Halmahera. Applied to plants and animals in nature, Darwin and Wallace independently realized, Malthusian population limits provided a means to generate new species from preexisting ones through the survival of individuals with beneficial variations. Wallace immediately set down his insight in a clear, tightly reasoned essay and sent it to Darwin, who had earlier expressed interest in Wallace's work. Wallace asked Darwin to review the manuscript and, if he thought it had merit, pass it along to Lyell, whom Wallace admired but did not know. Darwin was stunned by what he read.

Wallace's essay opened with a restatement of Malthus's population principle. "The life of wild animals is a struggle for existence," he asserted. For every species, far too many individuals are born than can survive, and each one of them is different. "As the individual existence of each animal depends upon itself, those that die must be the weakest," Wallace wrote in a passage that could have been drawn from his own hard life, "while those that prolong their existence can only be the most perfect in health and vigour." A beneficial variation could provide the edge needed for survival, he suggested, and, if so, it would be propagated in the survivor's offspring. Turning the most famous example of supposed Lamarckian evolution on its head, Wallace explained with emphasis, "The giraffe [did not] acquire its long neck by...constantly stretching its neck...but because any varieties which occurred among its antetypes with a longer neck than usual *at once secured a fresh range of pasture over the same ground as their shorter-necked companions, and on the first scarcity of food were thereby enabled to outlive them.*" Such variations, preserved and accumulated over time, would lead to new types, he reasoned. "Here, then, we have *progression and continued divergence* deduced from the general laws which regulate the existence of animals in a state of nature, and from the undisputed fact that varieties do frequently occur," Wallace concluded.[34] Darwin read these words as a precise summary of his theory.

Dejected, Darwin duly forwarded Wallace's manuscript to Lyell. "Your words have come true with a vengeance that I should be forestalled. You said this when I explained to you here very briefly my views of 'Natural Selection,' " Darwin wrote in his cover letter to Lyell. "I never saw a more striking coincidence. If Wallace had my manuscript sketch written out in 1842 he could not have made a better short abstract!... So all my originality, whatever it may amount to, will be smashed."[35] In his initial despair, Darwin somewhat exaggerated the identity between his well-worked theory and Wallace's burst of insight. Closer inspection would show that Darwin emphasized the role of individual competition in natural selection, for example, while Wallace stressed the selective power of ecological factors working on varieties. Lyell recognized the contributions of both men, and together with Hooker arranged for the Linnean Society of London to publish Wallace's essay jointly with two earlier writings in which Darwin had privately outlined his theory of natural selection. Darwin cooperated in this arrangement by supplying his writings; Wallace knew nothing about it at the time, but later expressed his satisfaction with it. The three items were read to the society at its meeting on July 1, 1858 (in alphabetical order by the author's last name), and Darwin promptly set about composing a more complete account of his theory. It appeared a year later as the book *On the Origin of Species by Means of Natural Selection, or the Preservation of Favoured Races in the Struggle for Life.* Darwin had preserved his priority in publishing the idea of the century.

CHAPTER 4

ENTHRONING
NATURALISM

By 1859, the idea of evolution did not seem as foreign or threatening as it once did to members of Britain's rising elite. Enriched by rapid industrialization at home and unprecedented colonial conquests abroad, they increasingly equated change with progress and saw their nation's economic and political ascendency as the natural consequence of its superior science and technology. Herbert Spencer's *laissez-faire* vision of social evolution became the popular philosophy of the age. The pastoral poetry of William Wordsworth, Britain's poet laureate until 1850, gave way to "Nature, red in tooth and claw" in the poetry of his successor, Alfred, Lord Tennyson. Then, on the day after Christmas, 1859, the staid voice of establishment England, *The Times* of London, featured a wonderfully favorable review of Charles Darwin's latest book, *On the Origin of Species*.

"There is a growing immensity in the speculations of science to which no human thing or thought of this day is comparable," the anonymous reviewer began. "Hence it is that from time to time we are startled and perplexed by theories that have no parallel in the contracted moral world." *Origin of Species* presents such a theory, the reviewer stated, "an hypothesis as vast as it is novel." Introducing readers to Darwin's view of life, the review hailed "the marvelous struggle for existence which is daily and hourly going on among living beings," killing far more individuals of every kind than could ever survive. "Those only escape which happen to be a little better fitted to resist destruction than those which die," the reviewer explained, and "their offspring will, by a parity of

reasoning, tend to predominate over their contemporaries."
By continuously selecting stronger varieties over time, this
process might propagate new species, the review stated, but
only further research could prove if it did. "This hypothesis,
may or may not be sustainable hereafter," the reviewer
stressed, "but its sufficiency must be tried by the tests of science *alone*... and by no others whatsoever."[1]

Delighted by this friendly review in the archconservative
Times, Darwin at first could only guess that its author was the
naturalist Thomas Henry Huxley. Huxley had not yet fully
accepted Darwin's theory of natural selection, but he enthusiastically embraced the criterion that scientists should invoke only natural causes to explain phenomena in nature. By
satisfying this fundamental standard, Huxley wrote, Darwin's
"ingenious" explanation for the origin of species had "an immense advantage over any of its predecessors." Empirical science, rather than revealed religion, must test its validity, he
reiterated. On this basis, the instinctively belligerent Huxley
would become Darwin's bulldog—actively taking the side of
evolution in public debate against all challengers. Every true
Briton should take up this cause on these terms, he urged in
his review, "if we are to maintain our position as the heirs of
Bacon and the acquitters of Galileo."[2]

Not yet thirty-five years old in 1859, Huxley had already
gained considerable power within the British scientific establishment, and his star was rising. From modest means and
with vocational training in medicine, he served as ship's surgeon aboard H.M.S. *Rattlesnake* during a Royal Navy surveying expedition to Australia from 1846 to 1850. Although the
ship carried an official naturalist, Huxley took it upon himself
to study marine invertebrates (particularly various types of
jellyfish, polyps, and mollusks) during the long voyage. His
detailed analysis of these little-known animals, published in
scientific articles that he sent back to England, opened doors

for him into the elite enclaves of British science upon his return—including election to the Royal Society of London in 1850, a permanent teaching position at the Government School of Mines beginning in 1854, and membership in numerous scientific associations.

At the time, the norms of British science teetered between those of traditionists such as Adam Sedgwick, who accepted divine action as the ultimate explanation for various natural phenomena, and modernists such as Charles Lyell, who sought only naturalistic, materialistic answers in science. With his irreverent manner, aggressive bearing, and magnetic personality, Huxley naturally assumed leadership of the Young Turks in every scientific institution he joined and (riding the wave of secular modernity then sweeping western Europe) carried them to cultural dominance within British science. He developed expertise in vertebrate anatomy, embryology, and paleontology during his academic career in England, but made his mark as a reformer of British scientific institutions and education. Admired by his allies and despised by his adversaries, Huxley had become the very personification of Victorian science by his death in 1895. "I am quite wild over Huxley," the American evolutionary philosopher John Fiske wrote to his wife in 1873, after he first met Huxley. "I never saw such magnificent eyes in my life.... And, by Jove, what a pleasure to meet such a clean-cut mind! It is like Saladin's sword which cut through the cushion."[3]

During the mid-1850s, Darwin identified Huxley as a key potential supporter, but held back from confiding in him until after publicly announcing the theory of natural selection in 1858. Ever the materialist, Huxley had rejected earlier theories of evolution because of their reliance on quasi-spiritual life forces as the source of directional change. Indeed, he denied the very notion of progressive organic development because it suggested design or purpose in nature, and (along

Thomas Henry Huxley, from a photograph taken in 1857, about the time that Darwin announced his theory of natural selection.

with Lyell) turned with vengeance against Richard Owen when the senior anatomist jumped to the progressivist camp. Owen's belief in designed archetypes guiding this progressive development only made matters worse, in Huxley's eyes.

Darwin knew that his theory of natural selection allowed evolution to proceed without purposeful life forms or divine design. Nevertheless, he worried that an inflexible opposition to evolutionism might keep Huxley from accepting the arguments in *Origin of Species,* and was overjoyed when it did not. Having finished the text on the very eve of its publication, Huxley wrote to Darwin, "You have earned the lasting gratitude of all thoughtful men. And as to the curs which will bark and yelp, you must recollect that some of your friends...are endowed with an amount of combativeness which...may stand you in good stead. I am sharpening my claws & beak in readiness."[4] In his letter, Huxley reserved judgment on several points in the book, but these reservations did not disturb

Darwin. "When I put pen to paper for this volume," he wrote back to Huxley, "I then fixed in my mind three judges, on whose decision I determined mentally to abide. The judges were Lyell, Hooker & yourself. It was this that made me so excessively anxious for your verdict. I am now content."[5] Huxley had finally joined Lyell and Hooker in endorsing Darwin's basic concept. Together with Asa Gray in America, they would sally forth as the "Four Musketeers" of Darwinism, while the chronically ill Darwin promoted his theory mostly from his home in Downe through an ongoing campaign of scientific research and writing.[6]

Expanding on his position in an 1860 book review, Huxley praised *Origin of Species* "as a veritable Whitworth [rapid-fire machine] gun in the armoury of liberalism"—the most effective new weapon for killing superstitious beliefs and clearing the field for rational materialism. "Old ladies, of both sexes, consider it a decidedly dangerous book," he wrote, but "all competent naturalists... acknowledge that [it] is a solid contribution to knowledge and inaugurates a new epoch in natural history." Huxley again held back from endorsing several points in Darwin's book and always maintained that "proof" of evolution by natural selection required producing new reproductively separate types through breeding experiments. Yet he accepted Darwin's hypothesis as "superior" to any earlier explanation for the origin of species—particularly the doctrine of special creation, which Huxley damned as "ow[ing] its existence very largely to the supposed necessity of making science accord with the Hebrew cosmogony" and as "a mere specious mask for our ignorance."[7] Privately to Darwin, Huxley vowed his willingness "to go to the stake, if requisite," in defense of several key chapters in *Origin of Species,* but more likely he viewed the book as a stake to drive through the heart of supernaturalism.[8]

Although he chaffed at Huxley's persistent depiction of

his concept of natural selection as an hypothesis rather than a theory, Darwin agreed that his book was "one long argument." More precisely, it presented two interwoven arguments: one contending *that* species evolve and another making the case for *how* they do. *Origin of Species* succeeded well enough in popularizing these arguments that, within four months of its publication, Huxley could hyperbolically claim, "Everybody has read Mr. Darwin's book, or, at least, has given an opinion upon its merits or demerits."[9] Almost overnight, one book transformed the scientific and popular debate over biological origins. A century and a half later, it still argues the case for evolution better than any other book.

———

Darwin began his argument with an analogy. Nineteenth-century Britons understood plant and animal breeding. Members of the aristocracy and landed gentry bred horses and hounds; small landowners cultivated hybrid plants in their gardens; farmers raised highly particularized strains of livestock and crops; even many urban workers bred pigeons. The opening chapter of *Origin of Species* examined how highly differentiated races of these plants and animals (particularly the lowly pigeon) are created and maintained through artificial selection. "The key is man's power of accumulative selection," Darwin explained. "Nature gives successive variations; man adds them up in certain directions useful to him. In this sense he may be said to make for himself useful breeds."[10]

Building on this analogy, Darwin then turned to wild plants and animals. In nature, careful observers find well-marked varieties within some species and nearly similar species within some genera, he noted, and fierce competition everywhere. "We behold the face of nature bright with gladness," Darwin wrote, "we forget that the birds which are idly singing round us mostly live on insects or seeds, and are thus constantly destroying life; or we forget how largely these songsters, or their eggs, or their nestlings, are destroyed by birds and beasts of prey." Here he invoked Malthus's principle of population: "A struggle for existence inevitably follows from the high rate at which all organic beings tend to increase," so that only the fittest competitors survive. Indeed, Darwin asserted, the fiercest struggle is among individuals of the same species "for they frequent the same districts, require the same food, and are exposed to the same dangers." He then asked, "Can we doubt (remembering that many more individuals are born than can possibly survive) that individuals having any advantage, however slight, over others, would have the best chance of surviving and of procreating their kind?...This preservation of favorable variations...I call Natural Selection."[11] Nature selects for fitness much like pigeon breeders select for fancy feathers, with both efforts propagating distinct varieties over time, Darwin argued. To him, well-marked varieties bled into incipient species—the process simply took longer.

When discussing natural selection, Darwin had in mind a survival-of-the-fittest process working on minute, inborn differences. Indeed, later editions of *Origin of Species* used the terms "survival of the fittest" and "natural selection" interchangeably. Supplementing natural selection, Darwin described a process of "sexual selection" whereby animals choose mates displaying attractive traits, such as the strength shown in a stag's rutting prowess or the beauty in a peacock's tail (the book only offered masculine examples). Darwin presented these traits, too, as innate and variations in them as minute. In his view, both types of selection worked on physical and mental traits, including animal instincts, to dictate which individuals left offspring.

Lacking an understanding of modern genetics and conceding that "our ignorance of the laws of variation is profound," Darwin speculated that innate differences among

individuals often came from environmentally caused distur-
bances in their parents' reproductive systems. He did not ex-
clude other causes, however, including variations simply
happening by random chance before birth. "Whatever the
cause may be of each slight difference in the offspring from
their parents—and a cause for each must exist—it is the
steady accumulation, through natural selection, of such differences, when beneficial to the individual, that gives rise to
all the more important modifications of structure," Darwin
concluded.[12] Asa Gray, who remained an orthodox Presbyterian despite his conversion to Darwinism, immediately seized
on this gap in Darwin's argument to propose that God guided
the evolutionary process by causing the beneficial variations
that selection acts upon in evolving new species. Over the
years, he developed this insight into a fully articulated theory
of theistic evolution, but Darwin rejected it. What sort of
God would use a survival-of-the-fittest mechanism to evolve
new species? Darwin effectively demurred. A Calvinist God
who predestined some for salvation and most for damnation
was Gray's obvious reply—which was hardly calculated to
gain a following in the modern era.

In the first edition of *Origin of Species* and increasingly so in
later ones, Darwin also conceded an evolutionary role for
variations acquired during an individual's lifetime. Adopting
a materialistic version of Lamarckism, Darwin included the
effect of use and disuse of organs, acclimatization to environmental changes, and correlated growth among body parts as
wholly naturalistic causes of organic change—and maintained that they, like inborn variations, were inheritable. Of
course, he did not understand heredity. "The laws governing
inheritance are quite unknown," Darwin admitted at the outset of his book. Yet he believed both that some variations
(whatever their cause) are inheritable and (quoting the Latin
maxim "*Natura non facit saltum*," or "Nature makes no jumps")

that all inheritable variations are, as he characterized them, "infinitesimally small." Attuned to discontinuities in the fossil record and perceiving no reason for limiting the size of variations, Huxley favored the so-called "saltationist" position that evolution proceeded in jumps through the inheritance of gross mutations. Clinging to his Lyellian intellectual heritage, however, Darwin always saw evolution as a gradual adaptive process. "As modern geology has almost banished such views as the excavation of a great valley by a single diluvial wave," he wrote in *Origin of Species,* "so will natural selection, if it be a true principle, banish the belief of the continued creation of new organic beings, or of any great and sudden modifications in their structure."[13]

Darwin envisioned the ongoing process of variation, competition, and selection generating a branching pattern of speciation. "Varieties are species in the process of formation," he wrote in *Origin of Species,* and "the modified descendants of any one species will succeed by so much the better as they become more diversified in structure, and are thus enabled to encroach on places occupied by other beings." He called it his "principle of divergence"—one species radiating into many, with each exploiting a different ecological niche. The end result of this pattern would look more like Owen's radiating archetypes than Lamarck's linear transmutation.[14]

Darwin first conceived of divergence as occurring on separate islands in an archipelago, where a single immigrant species might adapt differently to conditions on the various islands. By the time he wrote *Origin of Species,* he saw it happening everywhere, as the varied members of a species competed against one another for limited resources. Comparing the process to a growing tree, with the budding twigs as living species, he wrote, "at each period of growth all the growing twigs have tried to branch out on all sides, and to overtop and kill the surrounding twigs and branches, in the same manner as species and groups of species have tried to overmaster other species in the great battle for life." Just as only a few twigs survive to become branches, Darwin added, "so with the species which lived during long-past geological periods, very few now have living and modified descendants." At this point, he inserted the book's lone illustration—a sketchy time graph with a handful of lines starting at the bottom, some branching as they rise, and most ending before reaching the top. For Darwin, it represented the "Tree of Life."[15]

Darwin devoted the final chapters of his book to reciting examples of the effects of evolution in action. Of course, he could not claim that anyone ever observed a new species springing from an old one. The process took too long, Darwin asserted. Instead, he tried to show that known species relate to one another in ways that fit a radiating evolutionary pattern. Here was Darwin at his polemical best—presenting pointed evolutionary explanations for various familiar observations drawn from different branches of biology and geology. Progression in the fossil record; the existence of natural groups or families of species; the geographic proximity of similar species; anatomic, morphologic, and embryologic similarities among different species—all these made perfect sense under his theory, Darwin asserted, but were inexplicable under creationism. In a typical example near the end, he wryly noted that useless, rudimentary organs, which present "a strange difficulty . . . on the ordinary doctrine of creation, might even have been anticipated, and can be accounted for by" his theory.[16] By this point in his five-hundred-page book, Darwin was simply piling on evidence.

Origin of Species offered a new way of looking at life, and reached audiences far beyond the scientific community. It sold out its initial printing on the first day and was reissued in six revised English editions and eight foreign translations during Darwin's lifetime. Darwin closed his booklong argument with a rhetorical flourish. "It is interesting to contemplate an entangled bank," he wrote, "clothed with many plants of many kinds, with birds singing on the bushes, with various insects flitting about, and with worms crawling through the damp earth, and to reflect that these elaborately constructed forms, so different from each other, and dependent on each other in so complex a manner, have all been produced by laws acting around us."[17]

His final sentence, though, offered a bone to creationists. Something, he wrote (perhaps God, perhaps nature—he did not say), "breathed" life into the first living thing, or perhaps the first few living things—maybe the first plant and the first animal, or the first members of the most basic kinds of plants and animals (such as Cuvier's four *embranchements* of the animal kingdom). "From so simple a beginning," he concluded, all living things "have been, or are being, evolved."[18] The passage was thin gruel for creationists, but was too much for Huxley, who described it as "the very passage in Darwin's book to which, as he knows right well, I have always strongly objected." He thought Darwin should "give up the problem [of initial origins] or admit the necessity of spontaneous generation."[19]

———

Despite Darwin's closing concession, *Origin of Species* dealt a body blow to traditional Western religious thought. At a superficial level, Darwin's chronology for the origin of species differed on its face from that set forth in Genesis. Species evolved from preexisting species over vast periods of time, he asserted; God did not separately create all of them in a few days. Early-nineteenth-century theories of geologic history had already forced many Christians to stretch the Genesis account beyond its literal meaning, however, without any apparent crisis of faith. Even Lyell's uniformitarian geology, which all but eliminated the notion of any chronology for the earth's history, provoked little religious opposition.

Lyell's uniformitarianism, like the old-earth catastrophism of Cuvier and Sedgwick, did not challenge the doctrine of special creation in biology, however. If anything, it expanded the Creator's role by requiring ongoing or recurrent creation events. Darwin's theory dispensed with the need for a Creator to design species: Natural processes alone could produce each feature, trait, and instinct of every species. In an extreme example featured in *Origin of Species*, Darwin hypothesized how even the hive bee's instinctive ability to create perfectly hexagonal honeycombs (long seen as a testimony to divine design) could have evolved incrementally through nature selecting for efficient wax use. Although Darwin expressly left open the possibility of God creating the first living things, he pushed that event far back in time and out of the realm of science.

By replacing a divine Creator with a survival-of-the-fittest process as the immediate designer of species, Darwin's theory undermined natural theology. This carried cultural significance in the English-speaking world, where natural theology served as an organizing concept in science and an intellectual prop for Protestant Christianity. Both fields felt the impact. On the one hand, British and American naturalists tended to view living things through the lens of natural theology by attributing their fine adaptations to an all-knowing, all-loving Designer. In a single 1836 passage, for example, William Buckland described the powerful jaws and sharp teeth of his prehistoric *Megalosaurus* as "designed" both "for providing food to a carnivorous creature of enormous magnitude" and "to diminish the aggregate amount of animal suffering... [by being] adapted to effect the work of death most speedily."[20] Under Darwinian natural selection, in contrast, adaptations solely served their possessor's self-interest. On the other hand, at a time when rationalists pressed the case against biblical revelation, many Protestant theologians in Britain and the United States turned to the physical creation for seemingly objective evidence of the Creator's existence and character. In the most famous passage of this kind, Cambridge theologian William Paley in 1802 compared the complexity of a mechanical watch to that of a human eye. Just as the watch obviously had an intelligent designer, so must the eye, he wrote, because "every manifestation of design, which exists in the watch, exists in the works of nature."[21] Thus, for Paley, the eye proved the existence of a supernatural Designer, and its apparent perfection testified to the Designer's character. For Darwin, however, natural selection propagated better-adapted organs and organisms, not perfect ones. Absolute scales did not exist under Darwinism.

Darwin pressed the case for natural selection over special creation by citing examples of self-serving cruelty and lack of perfection in nature. Think of the ichneumon, whose "eggs are deposited in the living bodies of other insects," he wrote, or of introduced species displacing native ones on oceanic islands.[22] For nineteenth-century British and American Protestants weaned on natural theology, such passages made the book seem particularly (and perhaps intentionally) heretical. "I have read your book with more pain than pleasure," Sedgwick wrote sadly to Darwin within a week of receiving a prepublication copy. "'Tis the crown & glory of organic science that it *does* thro' *final cause,* link material to moral.... You have ignored this link; &, if I do not mistake your meaning, you have done your best in one of two pregnant cases to break it. Were it possible (which thank God it is not) to break it, humanity in my mind, would suffer a damage that might brutalize it."[23]

Writing to Darwin from America, Gray also expressed concern about the materialistic implications of *Origin of Species*. "I had no intention to write atheistically," Darwin replied early in 1860. "But I own that I cannot see as plainly as others do...evidence of design & beneficence on all sides of us. There seems to me too much misery in the world. I cannot persuade myself that a beneficent & omnipotent God would have designedly created the Ichneumonidae with the express intention of their feeding within the living bodies of Caterpillars." Turning to Paley's famous analogy, Darwin then added, "Not believing this, I see no necessity in the belief that the eye was expressly designed." Even human nature and mental ability might result from natural processes, he offered.[24] For those believing in natural theology, such reasoning represented the ultimate challenge of Darwinism. Beneficial variations were random and natural selection was cruel. If nature reflected the character of its Creator, then the God of a Darwinian world acted randomly and cruelly. The escape from this dark line of reasoning came in denying either natural theology, God, or Darwinism.

The sequence in Darwin's letter to Gray is telling. It passed quickly from observations of what seemed bad in nature (such as cruel animal behavior) to their implications for what seemed good in it (such as the human eye), and then moved on to ponder the origin of what seemed best of all, human morality and mentality. In *Origin of Species*, Darwin avoided making comments about human evolution, fearing that they would prejudice readers against his general theory, but his private correspondence (including his letter to Gray) showed his continuing fascination with the issue. Indeed, once his theory was successfully launched, Darwin published several books dealing with its implication for the development of human traits. This was what most intrigued others about Darwinism from the outset. Many early reviews dwelt on such matters. Even Sedgwick closed his 1859 letter to Darwin with a lighthearted reference to himself as "a son of a monkey." More than anything else from *Origin of Species*, "man's brute ancestry" became the topic of the day.[25]

———

While Darwin avoided commenting publically on human evolution, Huxley took up the cause and made it his own. It helped that Owen chose to attack Darwin's theory by exaggerating the anatomic differences between apes and humans—particularly between the structure of their brains—and thus to launch a preemptive strike against extending evolutionary ideas to cover man. Always eager to battle Owen, Huxley was soon writing and saying just the opposite. In 1863, he pulled together the pieces of his various arguments into a single popular polemic, *Evidence as to Man's Place in Nature*. "Whatever system of organs be studied," Huxley concluded, "the structural differences that separate Man from the Gorilla and the Chimpanzee are not so great as those which separate the Gorilla from the lower apes."[26] To illustrate his argument, the frontispiece showed a highly doctored sequence of rising primate skeletons, gibbon to man, walking in profile from left to right. One clearly led to the other, and ultimately to us. "It was inspired visual propaganda," historian Janet Browne concludes.[27]

Huxley's moment of greatest glory came in Oxford in 1860, at the British Association for the Advancement of Science's annual meeting, then the world's largest regular conclave of scientists and the science-minded. Midway through the weeklong meeting, Huxley publically clashed with Owen over similarities between the brains of gorillas and humans. Then, two days later, Oxford's scholarly Anglican bishop, Samuel Wilberforce, delivered a long speech raising scientific objections to Darwin's recently announced theory. Wilberforce's speech was doubly damned in Huxley's eyes. Not only had Owen coached Wilberforce, but here was a cleric opining on a matter of science.

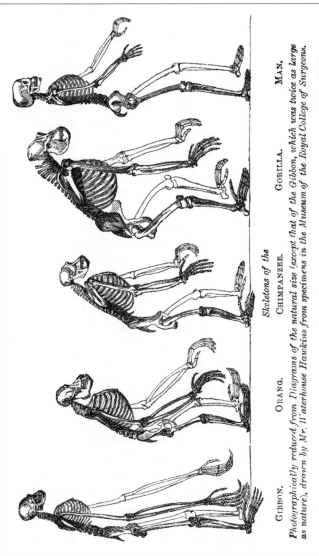

GIBBON.   ORANG.   CHIMPANZEE.   GORILLA.   MAN.

Skeletons of the

Photographically reduced from Diagrams of the natural size (except that of the Gibbon, which was twice as large as nature), drawn by Mr. Waterhouse Hawkins from specimens in the Museum of the Royal College of Surgeons.

T. H. Huxley's 1863 illustration emphasizing similarities in the skeletal structures of humans and other primates.

Huxley saw his chance near the end, when Wilberforce jok-
ingly asked whether Huxley was related to apes on his grandfather's or grandmother's side. "The Lord hath delivered him into
mine hands," Huxley whispered as he rose solemnly to reply.[28]
No transcript exists, but Huxley claimed to have said, "If then
the question is put to me would I rather have a miserable ape
for a grandfather or a man highly endowed by nature and possessed of great means and influence and yet who employs those
faculties for the mere purpose of introducing ridicule into a
grave scientific discussion—I unhesitatingly affirm my prefer-
ence for the ape."[29] The exchange quickly became legendary.
At a time when science and the Church battled for authority to
explain origins, Huxley had smitten the bishop in his own lair.
Those on Huxley's side felt the boundary line between science
and religion shift underfoot.

The questions raised by Darwinism that most vexed scientists and theologians concerned the origins of human mental
and moral attributes, particularly altruistic behavior. Could
these distinguishing human characteristics have evolved by a
naturalistic process, they asked, or did God create them, perhaps in an evolved human body? Traditionally, Christian theologians had attributed these attributes to an indwelling soul,
the existence of which lifted humans above other animals.
Scientists generally segregated humans from other animals
on this basis as well, from Aristotle's theory of the rational
soul found only in humans, through the Cartesian dualism
splitting physical matter from the human and divine soul, to
Cuvier's division of humans and primates into separate taxonomic orders. Conventional scholars had not looked toward
nature for an understanding of the human mind, human behavior, and human morality. Indeed, they studied humans on
their own terms through religion or moral philosophy. Now
Huxley, in *Man's Place in Nature,* lumped humans in the same
order with other primates and boldly asked, "Is the philan-

thropist or the saint to give up his endeavours to lead a noble life, because the simplest study of man's nature reveals, at its foundations, all the selfish passions and fierce appetites of the merest quadruped? Is mother-love vile because a hen shows it, or fidelity base because dogs possess it?"[30] These were the new questions of the Darwinian age.

———

After steering clear of the intense, often emotional public debate over human evolution for more than a decade, in 1871 Darwin finally articulated his thinking on the subject in a massive two-volume book, *The Descent of Man.* "The sole object of this work," he wrote, "is to consider, firstly, whether man, like every other species, is descended from some preexisting form; secondly, the manner of his development; and thirdly, the value of the differences between the so-called races of man."[31] The book did not have the impact of *Origin of Species.* Reviewers tended to dismiss it as the musing of a senior scientist, as well they might. "In many ways the book *was* the man," Darwin biographers Adrian Desmond and James Moore observe, "pudgy and comfortable, sedate in its seniority, full of anecdote and rather old-fashioned."[32] It featured tales of pet behavior and speculations about biologic origins of race and gender differences. Darwin's Victorian biases were apparent throughout. Indeed, like a Victorian curiosity shop, the book contained so many provocative details that its message was somewhat lost in the clutter. Nevertheless, it raised the key issues that would thereafter occupy researchers in the field.

Darwin's basic case for human evolution consisted of two main parts. First, he presented the by-then well-known evidence for the evolution of the human body. In anatomic structure and embryonic development, people resemble other animals, he noted, and the persistence of monkey-like rudimentary features (such as the tailbone) reinforces the conclusion that the human body evolved from lower forms. "To take any other view, is to admit that our own structure, and that of all the animals around us, is a mere snare laid to entrap our judgment," he wrote. Relying primarily on structural similarities, Darwin traced human ancestry from "the most ancient progenitors in the kingdom of the Vertebrata" through ancient fishes and amphibians, early marsupials and placental animals, to "the New World and Old World monkeys; and from the latter, at a remote period, [to] Man, the wonder and glory of the Universe."[33]

The body's evolution, even if accepted, did not settle the matter: Many believed that humans stood apart from animals because of their minds and emotions, not their bodies. Darwin thus extended his naturalistic analysis to include those mental and moral attributes that supposedly uplifted humanity, such as higher reasoning, self-consciousness, religious devotion, and the ability to love. The mental powers and moral feelings of humans differed in degree (rather than in kind) from those of other animals, he asserted, with a progressive gradient linking the lowest beasts to the highest humans. Darwin stressed the human-like qualities of higher animals (particularly pet dogs and wild monkeys) and the animal-like qualities of the "lowest" savages. "Can we feel sure that an old dog with an excellent memory...never reflects on his past pleasures in the chase? and this be a form of self-consciousness," he wrote in a typical passage. "On the other hand...how little can the hard-working wife of a degraded Australian savage...reflect on the nature of her own existence!" Similarly, Darwin doubted whether Fuegians felt religious devotion yet saw "some distinct approach to this state of mind in the deep love of a dog for his master." As for the vaunted power of human speech, Darwin wrote, "It does not appear altogether incredible, that some unusually wise ape-like animal should have thought of imitating the growl of a

beast of prey, so as to indicate to his fellow-monkeys the nature of the expected danger. And this would have been a first
step in the formation of language."³⁴

Darwin attributed the evolution of even the most ennobling of human traits to gradual, survival-of-the-fittest
processes. Long ago in Africa, he suggested, some anthropoidal apes descended from the trees, started walking erect in
the open spaces, began using their hands to hold or to hunt,
and developed their brains—all in incremental steps that
each helped to preserve the individual or its group. As in *Origin of Species,* the variations themselves were either inborn or
acquired, with beneficial ones propagated through natural selection. Darwin envisioned the winnowing process at work
among individuals, nationalities, races, and civilizations, with
plucky Englishmen (and their American scion) advancing to
the fore. "The remarkable success of the English as
colonists...has been ascribed to their 'daring and persistent
energy;' but who can say how the English gained their energy?" he asked. "Obscure as is the problem of the advance of
civilization, we can at least see that a nation [like England!]
which produced during a lengthened period the greatest
number of highly intellectual, energetic, brave, patriotic, and
benevolent men, would generally prevail over less favored
nations." Except perhaps in the most extreme climates, Darwin predicted the triumph of Europeans, whose highly developed cranial capacities he reported as 92.3 cubic inches on
average, as compared to 87.5 for native Americans, 87.1 for
Asians, and "only 81.9" for aboriginal Australians.³⁵ Here was
an argument rooted in its time and place.

Yet for all the power that Darwin attributed to natural selection in molding humanity and guiding its advance, he saw
the process as incapable of generating gender and external
racial differences. For those, Darwin resorted to and elabo-

rated on the secondary process of sexual selection that he had introduced in *Origin of Species* to account for the evolution of male mating traits, such as the peacock's tail.

Peoples differ in what they find attractive in mates, Darwin explained. Africans prefer dark skin and depressed noses while Europeans favor light skin and straight noses, he asserted, while "in Java, a yellow, not a white girl, is considered...a beauty." Within each race, Darwin speculated, sexual selection propagates and exaggerates favored external characteristics, as the most attractive mates are chosen first and have the most children. For instance, he proposed that Hottentot women have large bottoms because tribesmen desire mates who display that trait. Gender differences arise in a similar fashion, Darwin added, as "the strongest and most vigorous men...have been able to select the more attractive women" and together they "would have succeeded in rearing a greater average number of offspring." A thoroughly Victorian male, Darwin concluded "that the greater size, strength, courage, pugnacity, and even energy of man, in comparison with the same qualities in woman, were acquired during primeval times, and have subsequently been augmented, chiefly through the contests of rival males for the possessions of females." Similarly, he added, "As women have long been selected for beauty, it is not surprising that...women should have transmitted their beauty in somewhat higher degree to the female than to their male offspring." Applying similar reasoning, Darwin concluded that men possessed higher intellect; women displayed greater tenderness; and nature made it so.[36]

With *Descent of Man* and *The Expression of the Emotions in Man and Animals*, which appeared a year later, Darwin completed the account of evolution he began with *Origin of Species*. "In

Darwin's early anxious jotting such a story [of human evolution] seemed dangerously implausible," Desmond and Moore observe. "But now, habituated to material progress, social mobility, and imperial adventure, the *arriviste* reading classes lapped it up. A romantic pedigree suited them."[37] It certainly suited Darwin. Assuming that educated people already believed that the first humans were barbaric, he drew on earlier accounts in the book about monkeys and primitive peoples to conclude *Descent of Man* with a flourish: "For my own part, I would as soon be descended from that heroic little monkey, who braved his dreaded enemy in order to save the life of his keeper... as from a savage who delights to torture his enemies... and is haunted by the grossest superstitions."[38]

Descent of Man offered the first comprehensive naturalistic theory of human evolution, but it did not change many minds. Europeans and Americans had hotly debated the proposition that humans evolved from beasts ever since the publication of *Origin of Species* in 1859, but most continued to reject the idea long after the appearance of *Descent of Man* in 1871, including many evolutionists within Darwin's inner circle. For example, Alfred Russel Wallace, who remained a staunch Darwinist on other matters, became persuaded that an "Overruling Intelligence" created the first humans by ennobling anthropoidal apes with enlightened minds. "Natural selection could only have endowed the savage with a brain a little superior to that of an ape," he wrote in 1869 and maintained ever after, "whereas he actually possesses one but very little inferior to that of the average members of our learned societies."[39] Lyell promptly endorsed Wallace's position; Darwin considered them both traitors to his cause. For his part, Asa Gray steadfastly maintained that God supervised the beneficial variations that produced humankind. Even Darwin's bulldog, Huxley, envisioned evolution proceeding in jumps (rather than incremental steps) and believed that civilized humans could overcome nature in shaping their own destiny. Darwin had gotten his ideas on the table in the origins debate, however, and they remained the centerpiece of discussion thereafter. People would never look at themselves or each other quite the same again.

CHAPTER 5

ASCENT OF EVOLUTIONISM

By the 1870s, Darwin was an international celebrity. Even if people did not believe they descended from apes, they talked about it—and about Darwin. And for many of those who did believe, Darwin became a kind of a secular prophet or high priest. Secluded in his remote country home at Downe, perpetually ill or supposedly so according to some, Darwin played the part of hermit sage receiving favored guests on his own terms. "We have been rather overdone with Germans this week," Emma Darwin complained in a typical letter to her son Leonard. "Häckel came on Tuesday. He was very nice and hearty and affectionate, but he bellowed out his bad English in such a voice that he nearly deafened us. However that was nothing compared to yesterday when Professor Cohn (quite deaf) and his wife (very pleasing) and a Professor R. came to lunch—anything like the noise they made I never heard."[1] Other days brought other visitors—some received, some not. Total strangers, uninvited and unannounced, would peer from beyond the gate or be turned away by servants at the door. Surveying the scene, Huxley sent Darwin a sketch of a kneeling supplicant paying homage at the shrine of Pope Darwin.[2] Given their almost visceral contempt for Catholicism, both Huxley and Darwin surely enjoyed the irony.

The young American philosopher John Fiske was one of Darwin's luncheon guests late in 1873, and he left a lifelong disciple. "Darwin is the...gentlest of gentle old fellows," Fiske reported back to his wife in the United States. "His long white hair and enormous white beard make him very picturesque. And what is so delightful to see as that perfect frankness and guileless simplicity of manner which comes from a man having devoted his whole life to some great idea, without a thought of self?" Fiske then added a comment typical of many accounts: "I am afraid I shall never see him again, for his health is very bad and he had to make a special effort I think to see me today. Of all my days in England I prize today the most."[3] Darwin lived another decade, though, and gave countless more seemingly final audiences. He also wrote a staggering number of personal letters during the 1870s—some fifteen hundred a year, biographer Janet Browne estimates—most of them in response to letters from strangers asking about evolution, offering odd bits of scientific information, inquiring about his religious beliefs, or simply requesting his autograph. He ordered preprinted cards and a signature stamp to assist him in his responses, but apparently answered most of his correspondence with handwritten notes.[4]

In the public's eye, Darwin came to personify the idea of organic evolution. Even evolutionists who did not accept his theory of natural selection acknowledged him as their master. Fiske, for example, was an intellectual follower of Herbert Spencer and, like Spencer, inclined toward a Lamarckian view of organic progress—yet he called himself a Darwinist. Ernst Haeckel, whose visit to Downe Emma Darwin noted in her letter to Leonard, also had Lamarckian leanings but characterized his science as Darwinian. Indeed, by the 1870s the term "Darwinism" could mean either the general concept of descent with modification or the particular theory of evolution by natural selection, with the two ideas treated quite differently. Darwin argued for both in *Origin of Species,* but never conflated them. "Personally," he wrote with emphasis to Asa Gray in 1863, "I care much about Natural Selection; but that seems to me utterly unimportant compared to the question of *Creation* or *Modification*."[5]

During the 1860s and 1870s, as scientists raised increasing doubts about the sufficiency of selection theory, Darwin revised *Origin of Species* to add ever larger doses of the Lamarckian notion that acquired characteristics also feed evolution. Indeed, the biologists who clung to the belief that the natural selection of inborn variations could sustain the evolutionary process came to be known as "neo-Darwinians" to distinguish them from Darwin himself, who held less dogmatic views. Selection theory continued to lose ground in the decades following Darwin's death in 1883—so much so that, by the turn of the century, biologists were speaking of its eclipse or demise. The theory of evolution never faltered, though, becoming ever more widely accepted. To avoid confusion in tracing these developments, the terms "Darwinism" and "neo-Darwinism" apply here only to the specific theory that evolution proceeds through the natural selection of minute, random, inborn variations, while the terms "evolution" and "evolutionism" apply to the general concept of descent with modification.

WANING OF CREATIONISM IN SCIENCE

The doctrine of special creation had dominated Western biological thought for so long that few could have predicted how quickly it would fall from grace. In the United States, for example, virtually no naturalist publicly endorsed the idea of organic evolution prior to the publication of *Origin of Species* in 1859, yet a dozen years later the prominent American paleontologist Edward Drinker Cope concluded that intervening developments had placed "the hypothesis on the basis of ascertained fact."[6] Princeton University geologist-geographer

Arnold Guyot was perhaps the last working American naturalist of high professional standing who publicly challenged
evolutionism, but by the time of his death in 1884, even he acknowledged that all living things except humans may have
descended from a common ancestor.

The end result was much the same in Britain. There, Darwin, Huxley, and their allies effectively collaborated to take
over the scientific establishment, with the goal of enthroning
naturalism as the ideology of science and science as the main-
spring of modern society. At first they consciously sought to
minimize open scientific debate over Darwinism while systematically advancing the interests of biologists who utilized
an evolutionary approach. Working through an intimate
group of like-minded intellectuals known as the X Club,
Huxley and his friends managed to assume leadership roles in
many of Britain's leading scientific societies, place supporters
in prominent university and museum positions, and influence
the editorial policies of scientific journals. In 1869, they
founded the journal *Nature* as the mouthpiece of scientific
naturalism, and unabashedly promoted Darwinism in its
pages.

By the 1870s, transmutation had supplanted special creation in Britain as the accepted scientific explanation for the
origin of species. The Paley tradition of seeing design in nature, so deeply entrenched in British science, did not disappear overnight—but its followers were either marginalized or
persuaded to accept some theistic variant of evolution theory.
Huxley outmaneuvered Richard Owen at every turn, and
soon left the elder statesman of British anatomy publicly
honored but scientifically isolated. Owen ended his career espousing an ill-defined theory of "derivation," whereby
species mutated into new species, presumably along prede-
termined paths or in accord with designed archetypes. "So
successful was this takeover of the British scientific commu-

nity," historian Peter Bowler says about the X-Club putsch, "that by the 1880s its remaining opponents were claiming that Darwinism had become a blindly accepted dogma carefully shielded from any serious challenge."[7]

During this period, Darwinism also spread throughout the British Empire, taking root wherever an Anglo culture prevailed—particularly in the new scientific institutions of Australia, New Zealand, and Canada. It spread more slowly outside the English-speaking world, however, and had notably less impact in the Roman Catholic domains of southern Europe and Latin America. Cuvier's legacy kept evolutionism at bay in France for a generation, and when it did enter French science, it did so with a distinctly Lamarckian flavor.

In the years immediately following the publication of *Origin of Species,* the key battleground for Darwinism was Germany. That country, then unifying out of a multitude of separate states, already stood out as the preeminent center for the study of morphology (or plant and animal structure), physiology (or organic functions), cell theory, and virtually every other branch of laboratory biology. There, beginning in the 1860s, morphologist Ernst Haeckel, a political radical and scientific materialist, used his own pseudo-Darwinian theory of evolution as a battering ram against the entrenched metaphysical idealism of German biology. This rich national tradition embraced stasis and preordained archetypes in nature, as against Haeckel's vision of self-driven progress through natural processes. With a growing corps of disciples, he sought to understand living things according to evolutionary genealogies rather than archetypical patterns. Where Haeckel saw evolution proceeding through the accumulation of Lamarckian acquired characteristics selected for fitness in a Darwinian fashion, his contemporary August Weismann advanced a purer form of Darwinism that relied exclusively on the natural selection of inherited variations, with those variations based in an individual's hereditary "germ plasm." With Haeckel, Weismann, and their followers, Darwinism (born in Britain and nurtured in the United States) found a home in Germany.

———

Everywhere evolutionism took root, it held a similar appeal for scientists. With a theory of evolution, laboratory biologists and field naturalists could begin trying to explain the origins of living things (and perhaps of life itself) in terms of regular, rational, repeatable natural processes rather than divine fiat. By that time, this was what scientists did. Science had featured such activities before, but not to the exclusion of all else. Indeed, Isaac Newton, Johannes Kepler, William Gilbert, John Ray, and many other prior scientific luminaries freely mixed matters of supernatural and natural causation. In the shadow of the Enlightenment, though, scientists increasingly subscribed to the tenets of methodological naturalism. Doing anything else became an abrogation of their responsibilities as modern scientists. For theists like Asa Gray and, perhaps, Owen, evolution might simply represent the immediate or secondary cause of new species (with God remaining the Prime Mover); for materialists like Huxley and Haeckel, it surely served as the ultimate or final cause of life. For laboratory biologists and field naturalists in either camp, however, it was becoming the only acceptable scientific answer to the origins question.

From the outset of their public campaign, Darwin and Huxley stressed that the theory of special creation simply was not scientific. Gray made a similar point, summarized in his statement that the principal strength of evolution theory "appears on comparing it with the rival hypothesis of immediate creation, which neither explains or pretends to explain" anything in biology.[8] Gray's collaborator in devising a theory of theistic evolution, geologist George Frederick Wright, added that, in pursuing science, "we are to press known secondary causes as far as they will go in explaining facts. We are not to resort to an unknown (i.e., supernatural) cause for explanation of phenomena till the power of known causes has been exhausted. If we cease to observe this rule there is an end to all science and all sound sense."[9] Wright's defense of methodological naturalism in science is especially telling because he was an ordained minister with proto-fundamentalist leanings, president of the still-evangelical Oberlin College, and addressing a religious audience. Whatever the source, clearly naturalism had triumphed within science.

Among scientists, the most prominent late-nineteenth-century holdouts against evolutionism were adherents of German idealism rather than biblical literalism. Owen represented this viewpoint in Britain, of course, and Harvard zoologist Louis Agassiz did so in the United States. Both of these world-renowned naturalists based their arguments against evolution squarely on evidence from nature suggesting that species represented ideal types. For Agassiz, as for his mentor Georges Cuvier, individual organisms simply exhibited too much complexity in themselves, their relationships to one another, and their fit into the environment to change in any fundamental way. Indeed, he stressed that the entire community of organisms must remain in balance or cease to exist. "Pines have originated in forests, heathers in heaths, grasses in prairies, bees in hives, herring in schools, buffaloes in herds, men in nations," Agassiz wrote in his classic 1856 "Essay on Classification," otherwise they could not survive.[10] These were arguments against the theory of evolution, however, rather than for any alternative view—and they ultimately failed to counter the force of positive arguments in evolution's favor.

EMERGENCE OF EVOLUTIONARY SCIENCE

While Agassiz and other anti-evolutionists increasingly slipped
into the role of naysayers, evolutionists began reinterpreting
nature in light of their theory of origins and pursuing the rich
research agenda it offered. The best young researchers were
inevitably attracted to the field, and they in turn uncovered
additional evidence for evolution. In *Origin of Species,* for example, Darwin gave new meaning to rudimentary organs
(such as the tailbone in humans) and homologous correspon-
dences in comparative anatomy (like the five-fingered bone
structure of mammalian hands, paddles, and wings). Useless
organs and less-than-optimal homologies made perfect sense
as by-products of evolutionary development but little at all as
the artwork of an Intelligent Designer. Seizing on this evidence for evolution, comparative anatomists and evolutionary morphologists found an ever-increasing number of such
features throughout the animal kingdom, and then moved beyond Darwin by using them to investigate genealogic relationships among species in a bold effort to diagram the
evolutionary tree of life.

Interest in reconstructing the evolutionary ancestry of
living things also focused attention on modern species that
appeared to connect fundamentally different kinds of plants
or animals. Some evolutionists saw marine lancelets, which
lack bony structures, as living links between invertebrates and
vertebrates, for example, and lungfishes, which can breathe
air for short periods, as a bridge between fish and amphibians.
Similarly, modern monotremes, including the egg-laying
platypus and echidna, and marsupials, which bear underde-
veloped young, seemed to tie reptiles to mammals. Of course,
Agassiz and other critics countered that the persistence of
these and other ancient forms should argue against a univer-
salized law of evolution. Indeed, since antiquity, naturalists

had viewed such intermediate species as evidence for a created chain of being stretching in gradations from the simplest organisms to the most complex ones. Further, even under the theory of evolution, the fact that intermediate species now exist does not prove that they once served as genealogic links between species. Instead, they could have evolved over time into forms that simply look as though they served as bridges between other modern types. Nevertheless, some scientists and many within the general public saw living links as convincing evidence for evolution. Haeckel in particular relied on them in constructing highly speculative evolutionary genealogies, and invented the name "*Pithecanthropus*" (Greek for "apeman") to identify the hypothetical missing link between humans and living anthropoids.

More controversially, Haeckel marshaled the combined expertise of German microscopy and comparative embryology to hunt for evolutionary relationships among vertebrates. His thinking in doing so stemmed from his Lamarckian conception of evolution. If evolution proceeded through the accumulation of acquired characteristics added on to earlier types, Haeckel reasoned, then as each subsequent organism developed, it should first pass through its ancestral forms before taking on its evolved adult structure. He envisioned this "recapitulation" of past forms occurring in developing embryos, and examined them for clues to the evolutionary genealogies of their species. Recapitulation need not happen under Darwinism, where variation is posited to occur at conception (or at least before birth) rather than in later life, but even Darwin took note when Haeckel began finding evidence of it in the embryos of various vertebrate species. Haeckel's embryo sketches, drawn from a diverse array of animals ranging from fish and tortoises to hogs and humans, gave visible proof of evolution for all to see. His embryos looked virtually the same in their earliest stages, and only deviated as they de-

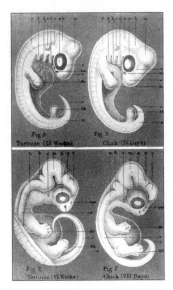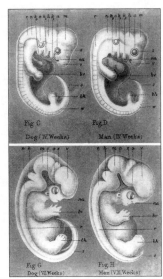

Ernst Haeckel's 1876 comparative sketches emphasizing initial similarities in the developing embryos of four vertebrate species: tortoise, chicken, dog, and human.

veloped. Ontology (or the developmental path of the individual) recapitulates phylogeny (or the evolutionary history of its group), Haeckel proclaimed. He called it his "biogenetic law." Opponents could only question his observations; his conclusions from these observations seemed irrefutable. As it turned out, Haeckel had greatly exaggerated the similarities of the early embryos. For a time, however, Haeckel's biogenetic law served as a powerful argument for evolution and his sketches were widely reprinted.

Haeckel's brand of evolutionism provided scientific support for the particularly virulent racism that infected some strains of German thought during the late nineteenth and early twentieth centuries, culminating in Nazi race theory.

Races, groups, and nationalities evolve in direct response to their environment, he urged, with humanity advancing through a competitive struggle for existence among them. Building on his view of the fundamental unity of the spiritual and the material, Haeckel formulated a secular philosophy of "monism" that championed a strong, centralized state as the driving force for human progress through racial competition, group sacrifice, and international war. The Monist League carried these ideas into politics by promoting German unification and expansion in the decades leading up to World War One. Monists enthusiastically supported the war effort and, following Germany's humiliating defeat in 1918 and Haeckel's death nine months later, some of them turned to Adolf Hilter's new National Socialist Party and its doctrines of racial, ethnic, and national superiority as vehicles to advance their ideals. The connection between evolutionary science and national socialism is highly complex and undoubtedly indirect, but nonetheless real.[11] For good or bad (and in this case for the very worst), scientific ideas can have social meaning. Haeckel's biology helped unleash the militant nationalism and murderous racism that cultural and social norms usually keep in check.

In his commitment to laboratory research, Haeckel represented a rising trend within the life sciences—but biological field research continued as well, and it, too, increasingly reflected an evolutionary perspective. Of course, Darwin and Wallace were first and foremost field naturalists, and their observations about the geographical distributions of native species in South America and the South Pacific had inspired their thinking about evolution. Both relied heavily on such evidence in making their cases for organic evolution.

Other naturalists followed these lines of investigation. Using emerging data regarding the geographical distribution of land birds, for example, Huxley's closest friend in science, the influential British ornithologist Philip Lutley Sclater, had already divided the earth into six zoologically distinct regions, each with its own characteristic avian populations. As originally conceived in 1858, his scheme conformed to the notion of zoogeographic regions tied to the special creation of species in places suited for them, but it gained new meaning in light of evolutionary theory. His regions perfectly fit Darwin's idea that the basic types of land animals evolved on the various large continental landmasses, and then became modified to fit local conditions as they spread. Physical barriers to distribution (primarily oceans, but also deserts and mountain ranges) coupled with past or present land bridges and island stepping-stones produced the distinctive zoogeographic regions and accounted for the absence of land mammals on oceanic islands. During the 1860s, Sclater (then secretary of the Zoological Society of London and one of the best-connected scientists in the English-speaking world) pressed his zoogeographic regions into the service of Darwinism. At the same time, Darwinian botanists Joseph Hooker at Kew's Royal Botanic Gardens and Asa Gray at Harvard University supplied evolutionary interpretations for the distribution patterns of plant species. This became some of the best scientific evidence for evolution.

For the half century from his return to Britain in 1862 to his death there in 1913, no one was more associated in the public mind with questions of biogeography than Alfred Russel Wallace. Throughout this period, he earned his living as a popular author, specializing in science writing and social commentary. His engaging style and association with Darwinism assured a market for his two dozen books and scores of articles, many of which dealt with biogeography—a topic that appealed to Victorian interests in exotic plants and animals. His 1869 *Malay Archipelago* discussed the dramatic line between South Asian and Australian species that splits the

East Indies. The two large islands of Bali and Lombok are less than fifteen miles apart and yet, as Wallace noted, "these islands differ far more from each other in their birds and quadrupeds than do England and Japan," both of which feature animals common to the Eurasian landmass.[12] This barrier became known ever after as "Wallace's line." In the landmark 1876 *Geographical Distribution of Animals,* Wallace extended Sclater's analysis of avian geography to animals in general, finding the same basic zoogeographic regions and reinforcing their scientific significance. Wallace's final book of this type, the 1880 *Island Life,* completed his global study of biogeography by examining the distribution of plants and animals on continental and oceanic islands, fitting it neatly into an evolutionary pattern.

In *Island Life,* Wallace elaborated on the significance of biogeography under an evolutionary view of life. The distribution of the various species provided critical clues for deciphering the earth's geologic, geographic, and climatological history, he asserted, as well as for charting the evolutionary genealogy of the species themselves. "A knowledge of the exact area occupied by a species or a group is a real portion of its natural history," Wallace noted. "We can never arrive at any trustworthy conclusions as to how the present state of the organic world was brought about until we have ascertained with some accuracy the general laws of the distribution of living things over the earth's surfaces."[13]

Wallace did not limit his popular writing to scientific issues in biogeography or evolution theory, but (like Haeckel) ranged freely over the social, political, and spiritual implications of evolutionism. Although both Wallace and Haeckel called themselves Darwinists, they found very different social meaning in their science. Wallace was a pacifist and agrarian socialist who earnestly believed that the answer to all social problems lay in the governmental redistribution of land and

wealth—a position he held before becoming an evolutionist
but that he later defended on evolutionary terms. "Surround
the poorest cottage with a spacious vegetable garden, with
fruit and shade trees, with room for pigs and poultry," he
wrote, "and the result invariably is untiring industry and
thrift, which soon raises the occupiers above poverty, and diminish, if they do not abolish, drunkenness and crime."[14]

Wallace (unlike Haeckel) did not attribute the emergence
of humanity and the rise of civilization to natural selection.
Instead, he credited a "spiritual influx" with endowing
evolved humans of every race and nationality with "reason,
the sense of beauty, the love of justice, the passion for truth,
[and] the aspiration toward a higher life."[15] These attributes
empowered humans to guide their subsequent evolution to an
ever higher plane, Wallace believed. In fact, he argued, "only
when we have so reorganized society as to *abolish* the cruel
and debasing struggle for existence and for wealth that now
prevails, shall we be enabled to liberate those beneficent natural forces which alone can elevate Character."[16] Chief in this
respect, he advocated economic and social equality for
women, which he envisioned as allowing them to choose lov-
ing mates rather than being forced into marriage with the
most powerful males. "In such a reformed society," he offered,
"the vicious man, the man of degraded taste or of feeble intellect, will have little chance of finding a wife, and his bad
qualities will die out with himself."[17]

Wallace also participated in the late-Victorian spiritualism
craze, which popularized (or at least publicized) efforts to
communicate with spirits of the dead. Indeed, his public defense of spiritualism helped persuade Darwin to attend one
seance, but the senior naturalist left early with the comment,
"The Lord have mercy on us all, if we have to believe in such
rubbish."[18]

Scientific ideas may have social meaning, but people supply the interpretation. The contrasting cases of Wallace and Haeckel give the lie to any simplistic conclusion about the social implications of evolutionary science. Evolutionary thought nurtured and sustained the former's egalitarian pacifism as readily it did the latter's proto-Nazi militarism.

DARWINISM'S STRUGGLE FOR EXISTENCE

The drift from orthodoxy of such professed Darwinists as Wallace and Haeckel shows that, by the end of the nineteenth century, Darwinism was on the ropes. Although Wallace relied more heavily on the natural selection of inborn variations to account for the normal course of evolution than even Darwin (who supplemented it with notions of acquired characteristics and correlated growth), he could not conceive of that process producing the great leaps forward represented by the first appearance of matter, life, animals, and humans. These steps required the intervention of an "Overruling Intelligence," he believed. Particularly, he viewed the human mind as so far superior to those of any other animals in ways not useful in the struggle for existence (such as moral reasoning and mathematical genius) that it could not have evolved in a Darwinian fashion. "Some of the greatest upholders of the theory of natural selection admit that these higher faculties cannot have been developed through its agency," Wallace asserted, citing Weismann and Huxley as examples.[19] He could have added Lyell and Gray, as well.

For his part, Haeckel maintained that humans evolved from "lower" forms in a purely materialistic manner just like other species, but viewed all evolution as guided by matter within the cell nucleus that somehow remembered acquired characteristics and passed them to future generations. This tilted him toward Lamarckism. Accordingly, like other Lamarckians, he tended to see evolution as a progressive, linear process.[20] Indeed, a widely reprinted illustration in his 1879 *Evolution of Man* pictured the evolutionary tree of life as a virtual lodgepole pine with humans at its apex, rather than the spreading beech of Darwinism, which branches into a broad, semispherical crown.[21]

One objection pushing biologists toward non-Darwinian versions of evolution involved the age of the earth. Inspired by Lyell's uniformitarian geology, Darwin originally assumed that natural selection had limitless time to grind out the present array of species. He did not know how long it would take, but envisioned the process as immensely slow. In 1866, the legendary British physicist William Thomson (later Lord Kelvin), who found the whole notion of Darwinism morally and scientifically repugnant, used his recognized expertise in thermodynamics to estimate the earth's age at about a hundred million years—or far less than Darwinism required. He derived this figure from the cooling time it should take for a newly formed earth-sized mass of molten matter to reach current terrestrial temperatures. Darwin acknowledged the force of Kelvin's calculation, but never fully accepted it. In response to it, however, many evolutionists looked for ways to accelerate the evolutionary process, such as by Lamarckian or theist factors. Not until the early twentieth century did physicists recognize that heat generated by the natural decay of radioactive elements greatly prolongs the earth's cooling process, and thus supplies added time for organic evolution. Kelvin's objections never stopped biologists from accepting an evolutionary view of life, but channeled them in non-Darwinian directions.

This channel was deepened by early attempts to solve the puzzle of inheritance. So long as species represented ideal, created forms, scientists could simply assume that those forms passed down through the generations, like begetting like, and dismiss individual variations as insignificant accidents of birth or development. Breeders could propagate varieties through artificial selection; yet let nature take its course, and future generations would revert to the species norm through random breeding. In contrast, for evolution to operate, variation must be a real, naturally sustainable attribute of the individual—and species must be simply clusters of similar, reproductively fertile individuals. Yet without fixed forms, how could hereditary information pass down through the generations?

If, as Darwin initially assumed, offspring inherited a blend of their parents' traits, then even the most beneficial variation in any one individual would eventually disappear through generations of breeding with normal types. Under any theory of blended inheritance, the Scottish polymath Fleeming Jenkin demonstrated, individual variations are "swamped" by the larger population. In an 1867 article, Jenkin predicted the fate of a European marooned on a tropical isle populated by Africans. "Our shipwrecked hero would probably become king; he would kill a great many blacks in the struggle for existence; he would have a great many wives and children," Jenkin hypothesized, using the imperialist assumptions of the day, "but can anyone believe that the whole island will gradually acquire a white, or even a yellow, population or that the islanders would acquire the energy, courage, ingenuity, patience, self-control, endurance, in virtue of which qualities our hero killed so many of their ancestors, and begot so many children; those qualities, in fact, which the struggle for existence would select, if it could select anything?"[22] Here lay all the elements of a dark Victorian novel: shipwreck, survival, sex, and swamping. For evolution to work, beneficial variations (whether large or small, inborn or acquired) must be fixed in individuals and passed to their descendants in some way, and (absent a working knowledge of genetics) Lamarckism seemed to offer the most plausible means.

In his massive two-volume 1868 *Variation of Animals and Plants Under Domestication,* Darwin proposed pangenesis as his solution to the inheritance puzzle. Under this hypothesis, each part of an organism generated tiny, unseen "gemmules" that carried hereditary information about itself. Gemmules existed for eyes and ears, for example, not for an entire organism, and all individuals inherited them at conception. Every ovum, sperm, and pollen grain contained "gemmules thrown off from each different unit throughout the body," Darwin proposed.[23] In the process of reproduction, gemmules for all parts of both parents passed to their offspring, where they combined to produce a unique new individual. With each individual inheriting two gemmules for every trait, the possible combinations were legion.

Although Darwin continued to believe that parental traits often blended in their offspring, he suggested in *Variation Under Domestication* that one gemmule could dominate and its counterpart lie dormant, perhaps to express itself in a later generation. This would permit some beneficial variations to persist without blending, and thus fuel the evolutionary process. Because gemmules came from every part of living organisms at the time of reproduction, moreover, they provided a material basis for transmitting acquired characteristics. If a giraffe had stretched its neck through use, for example, its gemmules could reflect this development and pass it to the next generation. By this time, Darwin welcomed ever more Lamarckism into his thinking as a way to speed the evolutionary process in response to Kelvin's estimate of the earth's age. Pangenesis permitted him to do so without admitting any of Lamarck's speculations about quasi-spiritual sources for acquired characteristics. Evolution could proceed in the face of Jenkin's and Kelvin's objections, and materialism still prevail.

Darwin's theory of pangenesis did not win many converts,

and other proposed solutions to the inheritance puzzle proved similarly unpersuasive. Many of them drew on Lamarckian or theistic forces because, so long as variations come from internal effort (as under Lamarckian evolution) or external direction (as under theistic evolution), then they could build within an entire population—and thus prevail. Random, individual variations were the most vulnerable to swamping—but they stood at the heart of Darwinism. True to form, Weismann and Wallace hued closest to Darwinism in addressing the puzzle of inheritance.

As a young microscopist in Germany, before his eyesight failed, Weismann became one of the first biologists to see the rodlike chromosomes that exist in the nucleus of every cell. Beginning in the 1880s, he theorized that these chromosomes consisted of "germ plasm," which purportedly carried hereditary information in a series of discrete germinal units. Like Darwin's gemmules, each germinal unit generated a particular body part, but, unlike gemmules, germ plasm for the whole body existed in every cell. In sexual reproduction, Weismann believed, germ plasm from both parents combined to produce their offspring's unique heredity, which thereafter remained fixed. Under his theory, inheritable variations in an individual's germ plasm occurred either at conception, when the parental germ plasm combined, or during a subsequent period of "germinal selection," when the germinal units from both parents competed in a struggle to determine which of them survived to express themselves in the individual. Variations in the germ plasm would persist in future generations without swamping, Weismann argued, but traits acquired after birth would die with the individual.

To discredit the concept of acquired characteristics, the pugnacious German conducted a polemic experiment in which he cut off the tails of baby mice for generations—without any visible shorting in hereditary tail length. At the time,

Lamarckians dismissed the experiment as irrelevant because it did not involve naturally acquired characteristics, but it became legendary after Lamarckism fell from favor. With germ plasm, Weismann found a mechanism of inheritance that fit a Darwinian model, except that selection occurred within an individual's germ plasm before birth as well as among individuals after birth. Due to its highly speculative nature and Weismann's dogmatic style, however, germ-plasm theory attracted only a small corps of very loyal followers.

Wallace did not go as far as Weismann in devising a Darwinian theory of inheritance, but he did recognize that swamping should not pose an insurmountable barrier if evolutionists conceived of variation occurring within populations rather than in individuals. Variations do not come in only two options, the field naturalist observed, but in a range of options centered over a hypothetical norm. If, in any group, more individuals survived at one end of the range than at the other, then the group's norm would shift—perhaps to form a new species. Population thinking required statistical analysis far beyond the capabilities of Wallace or Darwin, however, and only influenced evolutionary thought with the rise of biometrics around the turn of the twentieth century. Even at a conceptual level, though, Wallace could not believe that the range of variations within any subhuman population could generate the great leap represented by the appearance of the developed human mind. Here his faith in the power of natural selection broke down completely and he looked to the divine for help.

In addition to the age of the earth and the mechanisms of inheritance, other factors pointed evolutionists in non-Darwinian directions. The apparent persistence of gaps or discontinuities in the fossil record, for example, reinforced Huxley's saltationist position that evolution proceeds in jumps through gross mutations rather than through the incremental processes associated with Darwinism. Further, the continued absence of any known organic remains from pre-Cambrian strata bolstered the view held by Wallace and others that life did not appear gradually, but instead leaped into the scene. Perhaps most critically, many scientists continued to see a purposeful progression in organic history that seemed fundamentally at odds with the random directionlessness of Darwinism in general and the Darwinian view of variation in particular. Theistic evolutionists like Gray believed that God simply must guide the process. For some, it became a matter of profound moral or spiritual significance. British astronomer John Herschel (son of William) privately dismissed Darwinism as the "law of higgledy-piggledy." "What this exactly means I do not know," Darwin wrote to Lyell, "but it is evidently very contemptuous."[24] More to the point, although Herschel publicly conceded that evolutionary laws may serve as the immediate cause of life's development, he maintained that "an intelligence, guided by a purpose, must be continually in action to bias the directions of the steps of change—to regulate their amount—to limit their divergence—and to continue them in a definite course."[25] In saying so, he spoke for many.

NON-DARWINIAN THEORIES OF EVOLUTION

The general acceptance by European and American scientists of organic evolution, coupled with persistent doubts about the sufficiency of natural selection to explain it, left the field open for a flowering of other ideas. Among them, four basic approaches to the origin of species attracted the most attention from scientists: theistic evolution, Lamarckism, orthogenesis, and saltation (or mutation theory). These flourished side by side with Darwinism (or "neo-Darwinism," as Weismann's variant of it was called). Indeed, many scientists viewed the various approaches as complementary. Darwin himself supplemented natural selection with the inheritance of acquired characteristics and Wallace invoked spiritual influxes at critical junctures of the process. While multiple variants existed within each approach, they represent the broad diversity of evolutionary thought that followed in Darwin's wake.

In the United States during the late nineteenth century, Asa Gray virtually coopted the term "theistic evolution" for his theory that God guided the evolutionary process by supplying beneficial variations to species. In Britain, the Duke of Argyll and St. George Jackson Mivart separately devised alternative versions of theistic evolution in which a foreknowing God imparted direction into the laws of development themselves, so that species evolved over time to fit changed conditions. As an explanation for organic origins, however, theistic evolution failed the test of methodological naturalism that had come to define science. It had all but run its course as a serious scientific theory by 1900, and survived thereafter mostly as an ill-defined popular belief.

Weismann's tailed mice notwithstanding, the Lamarckian concept that characteristics acquired by use (or lost by disuse) could cause evolution retained a foothold within biology well into the twentieth century. By then, Lamarckism had spawned a related concept, known as "orthogenesis," which held that developmental trends, once ingrained in a species, would continue by their own internal momentum regardless of their adaptive value. Indeed, some Lamarckians used this concept to explain the extinction of species deemed to have overdeveloped features, such as the Irish elk, which supposedly had evolved antlers too large for its frame.

The German Lamarckian Theodor Eimer popularized orthogenesis during the 1890s through his efforts to explain the extreme, seemingly nonadaptive color variations of lizards and butterflies, but it found its largest following in an American school of self-proclaimed "neo-Lamarckians," which included such noted paleontologists as Alpheus Hyatt, Edward Drinker Cope, and Henry Fairfield Osborn. They freely invoked Lamarckian acquired characteristics and orthogenetic internal forces to explain the seemingly linear pattern of organic development that they detected in specimens from the rich fossil beds of the American West. The legacy of Louis Agassiz played a role here, too. He had trained a generation of American naturalists and, even though most of them came to accept evolution, many retained their teacher's anti-Darwinian bias. At a technical level, Agassiz doubted that evolution could operate with sufficient speed and direction to generate the diversity of life. Lamarckism and orthogenesis could account for both. They also gave a sense of purpose to natural developments, which was a hallmark of Agassiz's view of life. At theoretical and philosophical levels, Lamarckism and orthogenesis seemed to solve too many problems to be dismissed out of hand—yet biologists could never reliably document them happening in nature or in the laboratory. Support for both concepts evaporated rapidly once a plausible alternative appeared on the scene.

—

Saltation (or the theory that evolution proceeded in jumps fed by inheritable mutations) addressed many of the same problems as Lamarckism, but without the baggage of having to assume the unseen regarding the inheritance of acquired characteristics. It fit the available paleontologic evidence of discontinuities in the fossil record and relative rapid rate of organic development. Simply put, large variations seemed capable of driving evolution faster than small variations and, so long as enough individuals mutated to form a breeding population, they solved the swamping problem, as well. Best of all, researchers claimed to have documented cases of decidedly different varieties, subspecies, and even species appearing in a single generation, and thereafter breeding true to their new form. Dutch botanist Hugo de Vries led the way during the 1890s with his study of the evening primrose, *Oenothera lamarckiana,* which seemed able to sprout new, differently colored varieties at random. More than anyone else, de Vries transformed saltation into mutation theory, and in doing so pushed Darwinism near the verge of extinction as a viable scientific theory. De Vries himself retained a role for natural selection to pick the winners among competing mutations, but other mutationists thought selection was superfluous in this respect. For many young biologists, including William Bateson in Britain, Wilhelm Johannsen in Denmark, and Thomas Hunt Morgan in the United States, mutation theory offered a fresh, new alternative to tired old Darwinian and Lamarckian dogma.

In 1903, the German botanist Eberhard Dennert proclaimed, "We are now standing by the death-bed of Darwinism, and making ready to send the friends of the patient a little money to insure a decent burial of the remains." Conceding Dennert's verdict on Darwin's theory of natural selection, Stanford University entomologist Vernon Kellogg added in 1907, "It is also fair truth to say that no replacing hypothesis or theory of species-forming has been offered by the opponents of selection theory which has met with any general or even considered acceptance by naturalists. Mutations seem to be too few and far between; for orthogenesis we can discover no satisfactory mechanism; and the same is true for the Lamarckian theories of modification." By this time, theistic evolution did not even merit a nod among scientists. For Kellogg, Dennert, or virtually any other biologist, however, doubts about Darwinism and other mechanisms for forming species did not discredit the fact of evolution. "While many reputable biologists to-day strongly doubt the commonly reputed effectiveness of the Darwinian selection factors to explain descent," Kellogg asserted, "the descent of species is looked upon by biologists to be as proved a part of their science as gravitation is in the science of physics."[26] The challenge for biology became, How did evolution work?

CHAPTER 6

MISSING LINKS

The triumph of evolutionism within the Western scientific community did not translate into widespread popular acceptance of the theory, at least with respect to human origins (which was what people cared most about). Science did not then dominate how Europeans and Americans viewed the natural world—much less the supernatural. The matter of human origins was particularly sensitive because it impacted how people viewed themselves, other persons, and God. Crucially, evolutionary naturalism undermined belief in an indwelling spiritual soul, which for many people defined the very essence of humanness. The doctrine that God specially created the first people by giving them eternal souls carried with it certain implications about the meaning of human life; the theory that humans evolved naturally from soulless animals carried others; and many individuals found it difficult to switch. An 1871 cartoon in the British magazine *Punch* captured the tension. It showed an earnest young husband reading to his wife and infant child from Darwin's just-published *Descent of Man.* "So you see, Mary, baby is descended from a hairy quadruped, with pointed ears and a tail. We all are," he explained. His wife countered, "Speak for yourself, Jack. I'm not descended from anything of the kind, I beg to say; and baby takes after me."[1]

This captured the general mood. Although Darwin's *Descent of Man* titillated readers, it did not evoke the same high level of serious response as *Origin of Species,* which did not deal with human origins. Fewer magazines and newspapers reviewed the later book favorably, and more of them responded with humor, like the cartoon in *Punch*. At least one popular song, "The Darwinian Theory," ridiculed the notion that monkeys grew into men and a mass-produced figurine depicted a monkey sitting on Darwin's book and contemplating a human skull. Prominent evolutionary naturalists, such as Charles Lyell, Alfred Russel Wallace, and the Duke of Argyll, openly dissented from Darwin's view on human origins. Indeed, by 1908, Wallace could claim (with some hyperbole) that "all of the greatest writers and thinkers" agreed "that the higher mental and spiritual nature of man is not the mere animal nature advanced through survival of the fittest."[2] Novelist Leo Tolstoy proclaimed this viewpoint in Russia, for example, and prominent liberal protestant minister Henry Ward Beecher did so in the United States. Both embraced evolutionism to a point, but maintained that only God could make a soul. Roman Catholic Church doctrine fitfully gravitated toward accepting a similar position. During the late 1800s, even British prime minister William Gladstone made a point of endorsing the divine creation of humankind, as did the young American politician William Jennings Bryan.

Whether expressed in scientific commentary, church sermons, parlor discussions, or editorial cartoons, the basic sentiment was similar: Most people simply refused to believe their highly developed minds, morals, or emotions evolved from those of beasts. The gap appeared too great. They felt themselves superior to other animals. For many, these feelings carried more weight than an abstract scientific theory. "Really, Mr. Darwin," a stylish woman said to a tailed, apish-looking Darwin in an 1872 cartoon response to his *Expression of Emotions,* which immediately followed *Descent of Man,* "say what you like about man; but I wish you would leave my emotions alone!"[3]

Just as some people instinctively rejected the idea of human evolution, others embraced it for reasons that had little

Caricature of Charles Darwin appearing in 1872 following his publication of *The Descent of Man* and *Expressions of Emotions in Man and Animals.*

to do with science. Materialists, atheists, and radical secularists had long displayed a certain fondness for evolutionary theories of origins, such as Lamarckism—anything to dispense with God. Even though Darwin held strictly conventional political and economic views, his theory attracted the usual crowd. T. H. Huxley and Ernst Haeckel initially embraced Darwinism in part because it supported their anticlerical agendas for science and society. Karl Marx saw in Darwinism a scientific basis for his vision of class struggle. In 1873, he sent Darwin a copy of *Das Kapital* inscribed, "Mr. Charles Darwin on the part of his sincere admirer Karl Marx." An avowed capitalist, Darwin acknowledged the gift cordially, but never opened the book.[4] In America, feminist leader Elizabeth Cady Stanton welcomed Darwinism as a means to undermine what she saw as biblically based arguments for the subordination of women. "The real difficulty in woman's case is that the whole foundation of the Christian religion rests on her temptation and man's fall," she wrote in her 1895 *The Woman's Bible*. "If, however, we accept the Darwinian theory, that the race has been a gradual growth from the lower to a higher form of life, and the story of the fall is a myth, we can exonerate the snake, emancipate the woman, and reconstruct a more rational religion for the nineteenth century."[5]

From the conservative end of the political spectrum, the enormously influential social philosopher Herbert Spencer, already an evolutionist, freely worked Darwinian concepts into his progressivist philosophy of social development. As social theorists, Spencer and Darwin became inexorably linked in the public mind during the late nineteenth century. Spencer's many followers, whose numbers comprised a virtual social register of the British and American monied elite, typically embraced Darwinism as well. In his *Autobiography*, industrialist Andrew Carnegie recalled the day in the 1870s that his reading of Darwin's *Descent of Man* and various books by Spencer transformed his life. "I remember that light came as in a flood and all was clear. Not only had I got rid of theology and the supernatural, but I had found the truth of evolution," he wrote. "Man was not created with an instinct for his own degradation, but from the lower he had risen to the higher forms. Nor is there any conceivable end to his march to perfection."[6]

For people like Carnegie, Darwinism became a religion, or an alternative to religion. Pictorially, this sentiment appeared in a popular 1883 poster, attributed to London secularist George Holyoake, that purported to illustrate the fragmentation of the established British "National Church" into various factions ranging from High Church and Roman Catholicism to dissent and rationalism. In the upper-left corner, under the banner "Darwinism," an ape leads Spencer, Huxley, and other "agnostics" away from the central, umbrella-like dome of London's St. Paul's Cathedral toward a distant cloud of "Protoplasm." A bust of Darwin rises above the cloud. With his great white beard, Darwin could readily appear either God-like or apish—and, during the late nineteenth century, illustrators pictured him both ways.[7] It had little to do with science and lots to do with society.

———

Some evolutionists naturally sought to "prove" their view of life. At the time, the scientific evidence for evolution was largely circumstantial or hypodeductive. A wide variety of scientific observations about species (such as their geographic distribution, anatomic similarities, and natural groupings) supported an evolutionary interpretation of organic origins, but this evidence gained force primarily because of its cumulative effect. Each separate observation could have a nonevolutionary explanation. People predisposed against the theory could dismiss each bit of evidence separately, and then reject the whole or deny that it could account for human origins. Evolutionists lacked a critical experiment or irrefutable observation to prove evolution in a classic Baconian fashion. Huxley predicted that the production of new species through selective breeding would provide such proof, but it failed to happen as rapidly as Huxley hoped and never happened in clear-cut fashion capable of convincing skeptics. For many proponents of evolutionism, including Huxley, fossils became the most promising means to prove evolution. If paleontologists could uncover sequences of fossilized species leading to modern ones or linking different zoological groups (particularly humans to apes, but also the different classes of animals), then the public would believe the truth of evolution. This became the goal for a dedicated group of evolutionists.

Fossils had long served as both a basis for and barrier against belief in evolution. Georges Cuvier's path-breaking paleontologic studies had suggested that the species of any particular geologic time and place remained constant, and were replaced abruptly by an array of different forms. Human remains appeared only in relatively recent terrestrial deposits, he added, none of which predated the current period. At the time and ever after, these findings stood as evidence against evolution. In arguing for uniformitarianism in geology, Lyell countered that fossils are only laid down intermittently—when the conditions are ripe—so that discontinuities in the fossil record prove nothing. To him, the irregular appearance of ever more complex types of similar species suggested gradual succession over time rather than catastrophic disruptions.

Building on Lyell's argument, Darwin devoted two chapters in *Origin of Species* to showing that, despite notable gaps, the overall outline of the fossil record supported an evolutionary view of origins. In particular, he noted, the fossil record displayed a recognizable continuity in the succession of species within contiguous areas and a tendency toward greater organic variety and complexity over time. As unguided natural selection would suggest, he added, there was no fixed rate of change in the fossil record. Some organisms endured over many geologic epochs; others appeared and disappeared relatively rapidly; and not one ever reappeared once it became extinct. With further research, Darwin asserted, paleontologists would find many of the missing links in the evolutionary tree of life.

Throughout the late nineteenth century, paleontologists combed the fossil record for evidence of evolutionary development. Among their many finds, two stood out as particularly persuasive: fossils linking reptiles to birds and a sequence of fossils leading to the modern horse. Huxley played a part in both discoveries, along with American paleontologist O. C. Marsh. As Darwinists, both researchers interpreted these finds as supporting a branching (rather than linear) pattern of evolutionary development.

The links connecting reptiles to birds began turning up during the 1860s. At the time, some of the best-preserved fossils of the Jurassic period came from limestone quarries near the Bavarian town of Solnhofen. There, in 1861, workers found the fossilized remains of *Archaeopteryx,* a primitive bird with reptilian features. Although the specimen lacked a head, Huxley predicted that (like ancient reptiles, but unlike modern birds) the animal had a mouth with teeth. In 1872, Marsh identified two quite different species of toothed birds, *Ichthyornis dispar* and *Hesperornis regalis,* from Cretaceous-period fossil beds in Kansas and, five years later, a second specimen of *Archaeopteryx* turned up at Solnhofen, this one with a head and teeth. Here were reptile-like birds from the age of dinosaurs apparently evolving into a multitude of forms. The Solnhofen quarries also produced a small dinosaur, *Compsognathus lognipas,* which seemingly walked upright on birdlike hind legs and feet. In his 1868 paper, "On the Animals Which Are Most Nearly Intermediate Between Birds and Reptiles,"

Huxley presented *Archaeopteryx* and *Compsognathus* as two links in a chain connecting the modern classes of birds and reptiles.[8]

Fossil evidence of ancestral horses surfaced at roughly the same time. During the late 1850s, French paleontologist Albert Gaudry uncovered a three-toed type of prehistoric horse in fossil-rich Miocene deposits at Pikermi, Greece. This attracted the attention of evolutionists, who believed that modern single-toed horses must have developed from normal five-toed mammalian ancestors. By the early 1870s, Huxley and Russian paleontologist Vladimir Kovalevsky had fit Gaudry's find into a sequence of fossilized European horses stretching back to another three-toed type from the late Eocene. Marsh soon found an even richer array of ancient horses in the fossil beds of the western United States, including four and five-toed types from the early Eocene. Huxley hailed Marsh's sequence of horses as "demonstrative evidence of evolution."[9] In 1870, Darwin wrote to Marsh, "Your work on ... the many fossil animals of North America has afforded the best support to the theory of Evolution, which has appeared" since the publication of *Origin of Species*.[10] For his part, Marsh characterized his toothed birds and ancient horses as "the stepping-stones by which the evolutionist of to-day leads the doubting brother across the shallow remnant of the gulf, once thought impassable."[11]

Jurassic birds and Eocene horses simply represented the best-publicized and most dramatic paleontologic finds of the period. Other researchers identified an ever-increasing number of intermediate forms from the fossil record. Although such finds satisfied nearly all paleontologists that species evolved over time, they did not reveal how the process operated. Indeed, the apparent spurts and stops of evolutionary development displayed in the fossil record, coupled with the seemingly ordered succession of fossilized forms, led many researchers to favor Lamarckian or orthogenetic theories of evolution over Darwinian ones. Even Marsh came to believe that an internal force must propel brain growth across the generations.

Contrary to evolutionists' hopes, late-nineteenth-century paleontologic discoveries had little impact on public opinion. Critics still pointed to gaps in the fossil record as evidence against evolution. The continuing failure to find any fossils in pre-Cambrian strata suggested that complex forms of life abruptly appeared in the Cambrian. And the absence of fossils connecting humans to apes reinforced popular belief in the special creation of the first people. These gaps and missing links frustrated Huxley, Haeckel, and others philosophically committed to a wholly materialistic view of origins.

Despite cartoons and parlor jokes to the contrary, Darwin and most other evolutionists never claimed that humans evolved from modern apes. Rather, they asserted that all living primates, including humans and apes, had a common ancestor. This represented a radical departure from earlier scientific opinion. As if to emphasize the uniqueness of humans, for example, Cuvier and Owen had enthroned *Homo sapiens* in their own taxonomic order (and, in Owen's case, subclass), far removed from apes and monkeys. Differences in brain and hand structure served as the basis for this division. Asserting that Cuvier and Owen had grossly exaggerated these differences due to their anthropocentrism, Huxley (who assured readers in his 1863 *Man's Place in Nature* that he was "happily free from all real, or imaginary, personal interest in the results of the inquiry") argued that humans and apes belonged to a single order of primates.[12]

Placing humans and apes in the same order did not make them cousins, however. Cousins must have a common ancestor, and Huxley desperately wanted to find it. At the time, he had only two known types of hominid fossils to work with, both of European origin—the Engis skulls from Belgium, discovered in 1833; and Neanderthal bones from Germany, first uncovered in 1856, with more found later. Both types dated from an earlier geologic epoch, Huxley believed, but neither exhibited a brain size or structure sufficiently different from those of living humans to constitute a separate species. Of the Engis fossil, Huxley wrote, "It is, in fact, a fair average human skull, which might have belonged to a philosopher." As to the Neanderthal bones, he added, "In no sense [can they] be regarded as the remains of a human being intermediate between Men and Apes."[13] Somewhat different hominid fossils (together with prehistoric cave paintings) soon turned up in France, but these so-called Cro-Magnon creatures were apparently even more like modern humans than were Neanderthals.

In his 1863 *Antiquity of Man*, Lyell drew on such evidence to depict the first humans as appearing in a previous geologic epoch. This book stood in sharp contrast to his earlier *Principles of Geology*, in which he had agreed with Cuvier that humans did not predate the current epoch (although he agreed with him about little else). Lyell's new book related the story of humanity's gradual cultural development from the time of the Engis fossils, found amidst flint implements and the remains of extinct mammals, through the Neanderthal tribes, to the various living races.

The first scientific book of its kind written in an accessible style, Lyell's *Antiquity of Man* awakened public interest in human prehistory. Yet it held back from endorsing a Darwinian vision of human evolution. The human body may have evolved incrementally from lost types of anthromorphous apes, Lyell conceded, but the human intellect appeared the product of great leaps forward. "To say that such leaps constitute no interruption to the ordinary course of nature is more than we are warranted in affirming," he concluded.[14] Darwin felt betrayed. He complained to Huxley, "I am fearfully disappointed at Lyell's excessive caution."[15] Defending himself in a letter to Darwin, Lyell replied, "I have spoken out to the utmost extent of my tether, so far as my reason goes, and farther than my imagination and sentiment can follow."[16] Nevertheless, Lyell's book boosted the evolutionists' cause. It opened the way to ever-richer accounts of human cultural evolution that, for nearly half a century, substituted for any confirmed fossil evidence of human biologic evolution. Archeology and anthropology came of age during this period, breaking the stranglehold of historians, moral philosophers, and theologians over the study of early humans. Science was on the march.

———

Finally the breakthrough in protohuman paleontology came—from an unlikely place—because of the near-superhuman efforts of a driven Dutch Darwinist, Eugène Dubois. Born in 1858 into a respected, middle-class family in the conservative Catholic provinces of the southern Netherlands, Dubois rebelled against the staid traditionalism of his home region and cast his lot with science. As a student, he eagerly read Darwin, Lyell, and Huxley, but was particularly inspired by Haeckel's 1873 *History of Creation.* Haeckel opens with the challenge, "As a consequence of the Theory of Descent or Transmutation, we are now in a position to establish scientifically the groundwork of *a non-miraculous history of the development of the human race.* ... It follows from this theory that the human race, in the first phase, must be traced back to ape-like mammals."[17] This became Dubois's mission.

In 1881, when Dubois began his professional career as an assistant in anatomy at the University of Amsterdam, the theory of evolution was widely accepted within the northern European scientific community even if it had not yet made much headway among the general population. So far as biologists were then concerned, the best evidence for evolution came from highly technical studies of morphological relationships between species, and Dubois began climbing the academic ladder through such a study of the larynx. Speech distinguished humans from other animals, Haeckel stressed in *History of Creation,* so the development of the larynx should hold a key to human evolution.[18] Although Dubois excelled in this work, he was too restless and ambitious to be satisfied by it.

From his days as a freethinking youth in a traditional Catholic village, Dubois had fantasized about proving human evolution to everyone by finding a fossil link connecting humans to other primates, but he did not know where to look for it. In *Descent of Man,* Darwin suggested that humans evolved from African hominids.[19] Lyell was less certain. In *Antiquity of Man,* he noted that "anthropomorphous apes" live on the East Indian islands of Borneo and Sumatra as well as in tropical Africa, and urged naturalists to explore both places for the missing link in human evolution.[20] Given his extreme racial views, Ernst Haeckel had little difficulty choosing Asia over Africa as the cradle of humanity. In *History of Creation,* he hypothesized that a transitional form intermediate between apes and humans evolved on a lost continent located off the coast of South Asia. Building on early-Aryan mythology, Haeckel proposed that this creature's evolved descendants— at first upright, walking, apelike hominids (which he called *Pithecanthropus*) and later genuine, talking humans—spread across Asia and into Europe, where one branch developed into the Germanic race, which included Anglo-Saxons and the Dutch as well as modern Germans. This race, he wrote, "above all others, is in the present age spreading the network of its civilization over the whole globe, and laying the foundation for a new era of higher mental culture." A Lamarckian in his basic orientation, Haeckel believed that other human races sprang from less-developed branches off the same trunk or, perhaps, evolved separately from apes.[21] Dubois followed Haeckel's thinking about human evolution, and it led him to *Pithecanthropus.*

Fortuitously for Dubois, the Netherlands then ruled the East Indies as part of its colonial empire, and it was there that he decided to start his quest. No scientist had ever looked specifically for protohuman fossils. At the time, such matters were more a popular than a scientific concern—something that Jules Verne might write about in a novel. Funding was not available. Professional colleagues thought it ludicrous. Stubborn to a fault, Dubois quit his university post in 1887, signed on as a physician in the Dutch colonial army, and took his young family in search of *Pithecanthropus* on the Dutch East Indian islands of Sumatra and Java. Most remarkable of all, he found it after nearly four arduous years of looking.

Finagling time and support for his project once in the Indies and battling malaria along the way, Dubois examined caves, highlands, and riverbanks for hominid fossils. In 1891, his workers unearthed a protohuman molar and skullcap from an ancient stream bed near the tiny Javanese village of Trinil—the first such fossils ever found. A human-like thighbone turned up a year later about fifty feet upstream. Pieced together, Dubois named his discovery *Pithecanthropus erectus,* or "Upright Apeman," though most people simply called it "Java Man." The size of its braincase was intermediate between those of humans and apes, but the thigh was clearly made for walking upright, like a modern human's. This combination of traits fit Darwin's prediction that walking upright came first in the evolution of humans, with brain growth coming later. The location fit Haeckel's vision of Aryan origins. "The factual evidence is now in hand that, as some have already suspected, the East Indies was the cradle of mankind," Dubois boasted.[22]

Dubois's discovery created an international sensation. For decades, scientists debated Java Man's place in the evolution-

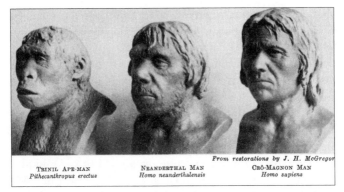

TRINIL APE-MAN
Pithecanthropus erectus

NEANDERTHAL MAN
Homo neanderthalensis

From restorations by J. H. McGregor
CRÔ-MAGNON MAN
Homo sapiens

A 1921 American Museum of Natural History exhibit showing three extinct hominid types, *Pithecanthropus,* Neanderthal, and Cro-Magnon, presented as a progressive series.

ary tree. Some saw it as simply a new species of gibbon; others considered it fully human. A few openly doubted that the skullcap and thighbone came from the same specimen. The discovery of a supposedly big-brained protohuman in 1912 at Piltdown, England, threw many researchers off the track by suggesting that brain size (rather than an upright gait) led the way in human evolution. Dubois doggedly maintained that *Pithecanthropus* represented an intermediate stage between humans and apes, in part by invoking his concept (derived from a curious mix of Lamarckism and mutation theory) that brain size doubled in each succeeding evolutionary step. By fudging the numbers, he calculated that *Pithecanthropus* had twice the brain capacity of apes and half that of humans—a perfect fit for the sole connecting link between the two types.

Returning to the Netherlands in 1895, Dubois regained a university post but never relinquished dictatorial control over his fossils. He withdrew them from examination by critics and grew increasingly paranoid. Yet his basic claim that *Pithecan-*

thropus was a direct ancestor of modern humans slowly gained acceptance. Beginning in 1929 with the discovery of similar fossils in China (popularly known as "Peking Man"), *Pithecanthropus*-like fossils turned up with increasing regularity across eastern Asia. Dozens more of the original type were found in Java. Paleontologists could no longer deny that the protohuman skull went with the more fully human thigh. These confirming discoveries gradually sidelined the Piltdown fossils even before they were discredited as a hoax in 1953. With further study of more specimens, paleontologists came to see Java Man and Peking Man as older and younger varieties of a single species much closer to modern humans than to apes, and rechristened them as *Homo erectus*. Researchers would have to look further back into the fossil record for the missing link between humans and apes. Aryan mythology notwithstanding, the trail led to Africa.

———

Late in the summer of 1924, a South African university student brought a fossilized skull to her anatomy professor, Raymond Dart. He identified the skull as coming from an ancient baboon and promptly sought more specimens from the source of the find, a limestone quarry at Taung. Two crates of fossils arrived later that fall. "As soon as I removed the lid a thrill of excitement shot through me. On the very top of the rock heap was what was undoubtedly an endocranial cast or mold of the interior of a skull," Dart later recalled. "I knew at a glance that what lay in my hand was no ordinary anthropoidal brain."[23] Hardly larger than that of a modern gorilla, this brain's shape was distinctly more human than that of any living anthropoid and its orientation suggested that the creature walked upright.

Dart rushed into print with his discovery. "Unlike *Pithecanthropus*, it does not represent an ape-like man, a caricature of a precocious hominid failure, but a creature well advanced

beyond modern anthropoids in just those characteristics, fa-
cial and cerebral, which are to be anticipated in an extinct
link between man and his simian ancestor," Dart announced
in the February 7, 1925, issue of *Nature.* "At the same time, it
is equally evident that a creature with anthropoid brain capacity ... is no true man. It is logically regarded as a man-like
ape." He called it *Australopithecus africanus,* "thus vindicating
the Darwinian claim that Africa would prove to be the cradle
of mankind."[24] A week later, *Nature* carried articles by
Britain's four leading anthropologists, all dismissing Dart's
claims as the rash assertions of a colonial naturalist. Dart may
have found a new species of anthropoid, they conceded, but
not a hominid. The Scottish-born anthropologist Robert
Broom, then living in South Africa, rushed to Dart's defense.
"In *Australopithecus,*" he replied in an article also published by
Nature, "we have a connecting link between the higher apes
and one of the lowest human types. ... While nearer to the anthropoid ape than man, it seems to be the forerunner of ... the
earliest human variety."[25] The debate raged on for years.

Dart made an observation in his initial article that clearly
set established anthropologists against him. In focusing their
search for early hominids on the tropics, he wrote, they had
been looking in the wrong place all along. Luxuriant forests
offer a comfortable home for anthropoids, Dart explained,
but "for the production of man a different apprenticeship was
needed to sharpen the wits and quicken the higher manifestations of intellect—a more open veldt country where competition was keener between swiftness and stealth, and where
adroitness of thinking and movement played a ponderating
rôle in the preservation of species."[26] The savannahs of
Africa, and not its jungles, nurtured humanity.

Although Dart never wavered in his belief that *Australopithecus* represented a link in human evolution, Broom became the idea's champion. A respected expert in African

fossils, he lobbied for *Australopithecus*'s place in hominid pale-ontology and actively searched for more specimens. In 1936, he found some in a cave at Sterkfontein, South Africa, and later found more. The new specimens established that *Australopithecus* walked erect and fit into the hominid line, either as an ancestor to modern humans or a branch that became extinct. Broom's fossils comprised two types. Some looked like Dart's original ones, *A. africanus,* while others displayed more "robust" development and were named *A. robustus.* Two decades later, in 1959, Kenyan paleontologists Louis and Mary Leakey discovered *A. boisei,* an even more robustly de-veloped species of *Australopithecus,* in the Great Rift Valley of East Africa, which runs through Ethiopia, Kenya, and Tanza-nia. "Lucy," a remarkably complete specimen of an appar-ently older type, *A. afarensis,* was discovered in Ethiopia during the 1970s by a team of researchers co-led by Ameri-can paleontologist Donald Johanson. Subsequent expeditions to the Rift Valley have uncovered fossils assigned to several more species of *Australopithecus*—some older, some younger, and some contemporaneous with the previously known types. What once looked like a virtually linear path of hominid progress (from *A. africanus* through *Homo erectus* to *Homo sapi-ens*) became an increasingly complex and seemingly branch-ing pattern of evolutionary development.

During the late twentieth century and into the early twenty-first century, paleontologists continued to find new types of hominid fossils in East and Central Africa. In 1961, the Leakeys identified a new human species, *Homo habilis* (or "Handy Man"), from fossil fragments discovered by their el-dest son, Jonathan. The species supposedly predated and gave birth to *Homo erectus.* Another son, Richard, confirmed the ex-istence of this earlier human type by finding further speci-mens of it during the 1970s. The Leakeys used newly developed radiometric-dating techniques to bolster their claim that the relatively big-brained, tool-using *Homo habilis* evolved in a land still inhabited by later forms of *Australopithecus,* but that these prior inhabitants ultimately became extinct. Using modern dating technology, paleontologists generally estimate that the various forms of *Australopithecus* lived from 4 to 1.5 million years ago, while the first human appeared about 2 million years ago, with *Homo sapiens* coming along only in the past three or four hundred thousand years. During the 1990s and after, paleontologists found even earlier hominid fossils (assigned to the genera *Ardipithecus, Orrorin,* and *Sahelanthropus*), with the earliest of these dated as up to 7 million years old. All of these hominids supposedly walked erect, and none are seen as a common ancestor connecting humans with apes. This would have happened even earlier.

The evolutionary tree for hominids is now as complete as the tree for any type of animal, and it fits a branching, Darwinian pattern. Upright posture apparently came first, presumably because it had survival value in an environment with mixed trees and grassland, then came bigger brains and tool use. Each new hominid fossil discovery generates front-page news around the world. Modern humans remain fascinated by their earliest ancestors. As Dubois predicted, hominid fossils now serve as the best-known and most widely accepted evidence for evolution.

CHAPTER 7

GENETICS ENTERS
THE PICTURE

Francis Galton believed that some people are born smart, others are born stupid, and the roots of all such hereditary traits reach deep into one's ancestry. He came to these beliefs from personal experience; his class and racial biases; the inspiration of the evolutionary theories of his first cousin, Charles Darwin; and what he viewed as his own hereditary genius. One of the last significant English gentleman scientists without institutional affiliation, Galton helped lay the foundation for both the modern science of genetics and the equally modern pseudoscience of eugenics. In this account, genetics comes first because it became central to the revival of Darwinian theories of evolution during the twentieth century. Eugenics will follow in due course.

During the mid-1800s, nearly all naturalists (including Darwin) accepted a blending view of inheritance whereby children manifest a middling mix of their parents' traits. Most of them also maintained that at least some characteristics acquired by individuals during their lifetimes passed to their descendants. Darwin came to regard acquired characteristics and other environmentally induced alterations in a parent's hereditary material as the major source for the inherited variations that fueled the evolutionary process. Although he was not a trained naturalist, Galton rejected both concepts and planted the seeds for their eventual overthrow.

Born in 1822 and freed from any obligation to earn a living after receiving an enormous inheritance upon his father's death in 1844, Galton first gained fame during the 1850s as a traveler and explorer. Adventures with Cambridge college chums in the Middle East gave way to serious expeditions through regions of southwest Africa never before traversed by Europeans. His technical and popular accounts of his explorations made him well-known in Britain by the time his cousin published *Origin of Species* in 1859. That book, Galton later claimed, changed his life. "I used to be wretched under the weight of the old fashioned 'arguments from design,'" Galton wrote to Darwin. "Your book drove away the constraint of my old superstition as if it had been a nightmare and was the first to give me freedom of thought."[1] Liberated from Christian mores and driven to promote the progress of civilization as he saw it, Galton turned to improving humanity by championing a purportedly Darwinian process of selective human reproduction, which he called "eugenics."

Based in part on his experiences in Africa, Galton believed that (on average) blacks were naturally inferior to whites in intelligence and other hereditary traits fitted to civilized life. Other nonwhite races fared little better in his estimation, and some even worse. Environmental factors cannot account for these racial differences, Galton asserted. Blacks in Europe or raised in white families remain much like their savage ancestors, while whites living in Africa maintain their civilized superiority, Galton explained, citing as authority the accounts of whites.[2] These observations convinced him that individuals cannot acquire heritable attributes through nurture or other environmental factors. Galton applied similar hereditarian reasoning to those he viewed as superior and inferior within a race, which fit his preexisting dismissive attitude toward welfare programs designed to uplift the underclass.[3] Reading (or misreading) *Origin of Species* apparently supplied him with a seemingly scientific basis for his societal views, and launched him on a midlife career in science to establish their validity. Surprisingly, he made some discoveries of lasting significance.

Galton's key insight, which he belabored in his many books and articles, was the concept of "hard" heredity, or heredity that cannot change during one's lifetime, as opposed to the "soft" heredity of heritable acquired characteristics. In his earliest published work on heredity, the 1865 magazine article "Hereditary Character and Talent," Galton rhetorically asked, "Will our children be born with more virtuous dispositions, if we ourselves have acquired virtuous habits? Or are we no more than passive transmitters of a nature we have received, and which we have no power to modify?" "No" and "yes" were the correct answers in Galton's mind. "We shall therefore take an approximately correct view of the origin of our life, if we consider our own embryos to have sprung immediately from those embryos whence our parents were developed, and these from the embryos of *their* parents, and so on for ever," he asserted.[4] For Galton, the body serves essentially as a passive receptacle and replicator for transmitting hereditary information across the generations rather than as an interactive participant in the process.

Seeking a material basis for this hereditary information, Galton commandeered Darwin's theory of gemmules and turned it to his own ends. For Darwin, gemmules represented invisibly small bits of hereditary information collected from every part of both parents and transmitted during reproduction to their offspring. New hereditary variations came from environmentally caused changes in gemmules as broadcast by a parent or germinated in the offspring. Galton accepted the basic idea of particulate gemmules, which he came to call "germs," but thought they remained distinct within the body, guiding its development yet never impacted by it. In reproduction, unmodified germs from both parents combine to compose the germs of their offspring, he asserted. Although all individuals (except identical twins) therefore carried their own unique mix of germs, Galton viewed such variations as merely a normal aspect of heredity as expressed in a population. Indeed, a born quantifier, he offered mathematical evidence suggesting that the distribution within a population of any given hereditary trait (which for him included complex characteristics like intelligence as well as such simple ones as height) fits a statistical bell curve, with most individuals falling at or near the species or racial mean and ever fewer toward the limits of variation in either direction. Most people are near the mean height for their race and gender, for example, but only a few greatly shorter or taller than it.

As long as no new hereditary information enters the equation, such as by acquired characteristics (which he rejected) or mutations (which he did not), Galton asserted that normal variations in the population cannot lead to the evolution of new species. Developing new techniques in statistics to prove his point, he calculated that successive generations descending from even the most gifted individual will naturally regress to the species or racial norm. Sustained selection for the same characteristic in both parents (such as can result from environmental pressures or eugenic mating) can slow the regression, but Galton purported to demonstrate that even it cannot permanently alter species or racial norms. Only sudden mutations (or saltations) can overcome the limits of ancestral inheritance and sustain the evolution of new species or races, he concluded. Setting aside his pessimism about the cumulative effect of sustained selection, Galton had hit upon many basic elements in the future synthesis of modern genetics and Darwinian theory—particularly the statistical interpretation of hard heredity.

Galton's idiosyncratic jumble of revolutionary insights and reactionary ideology inspired a generation of scientists and pseudoscientists. In Germany, cytologist August Weismann took up the cause of hard heredity—at first independently from Galton's work but increasingly with reference to it. Weismann called the heredity material "germ plasm," which quickly became its favored name, and localized it on chromosomes of each cell. In Britain, mathematician Karl Pearson and marine zoologist W.F.R. Weldon carried on Galton's statistical analysis of the cumulative impact over time of normal variations within populations—a field of study that became known as biometry. Although they did not win many converts, by 1900 Pearson and Weldon had demonstrated to their own satisfaction that Galton was wrong on one point: Sustained selection of continuous, normal variations (such as larger crabs over smaller crabs, in Weldon's experiment) could permanently shift the species norm in the selected-for direction. Over generations, they argued, this process could spawn new species without mutations. Concurrently, Galton's other principal British disciple, morphologist William Bateson, championed the role of mutations in fueling the evolutionary process and defended the position that normal variations lead nowhere. Although Pearson and Bateson clashed mightily over whether continuous or discontinuous variations fed evolution, both upheld Galton's position on hard heredity. By the turn of the century, Bateson was far from alone in accepting a saltationist theory of evolution. Disillusioned by the lack of solid evidence for either Darwinism or Lamarckism, an increasing number of biologists turned in the saltationist direction for answers, with Dutch botanist Hugo de Vries finding some of the most encouraging results.

———

Fields of finely hybridized flowers grow in the Netherlands. Working with such flowers in his research garden at the University of Amsterdam, de Vries detected any number of seemingly sustainable, inborn mutations—plants with new flower colors, stem shapes, or other distinctive traits sprouting out of seeds from known types. The most notable examples came from the evening primrose, but there were many others. It appeared almost as if one species gave birth to another in a single jump. After studying the phenomenon for several generations with the statistical techniques pioneered by Galton, de Vries developed a comprehensive theory of evolution by mutation to explain it. He proposed that, under certain stressful conditions, a sufficient number of mutants can appear within a single generation to create a sustainable breeding population. De Vries thought he had discovered the secret to the origin of species, and for a season many other biologists thought so, too. It turned out that his seemingly new flowers represented normal variations of highly complex hybrid types rather than new forms. In the course of his research, however, de Vries did uncover something revolutionary: Mendelian genetics.

De Vries began his study of plant heredity during the mid-1880s as a self-proclaimed Darwinist who nevertheless subscribed to a very hard view of heredity. Following Weismann, he attempted to update Darwinism with the latest developments in cell theory by proposing that hereditary information is carried in discrete units on chromosomes. By this time, microscopists had located rodlike chromosomes within the cells; seen them duplicate themselves in the normal process of cell division, with a complete set going to each new cell; and recognized that only half the normal number go into egg and sperm cells, so that fertilization restores the full complement by uniting half from each parent. Weismann, de Vries, and others inferred that these intricate processes must serve as the physical basis for heredity. Further, since there are more character traits than chromosomes, they surmised that hereditary information is carried on macromolecules situated on chromosomes rather than in the chromosomes themselves. De Vries called these hereditary units "pangenes" in honor of Darwin's theory of pangenesis. The concept of material pangenes was central to all de Vries's subsequent thinking about heredity, just as hypothesizing such entities to explain his observations was typical of how de Vries did science. "To climb up from the facts to a clear view of the general laws of nature," he declared in the inaugural address for his professorship, "is the goal of science."[5] He invariably claimed to see the big picture.

Since they were hypothetical, de Vries could tailor pangenes to serve his purposes. He proposed that normal variations for any particular trait (such as incremental difference in plant height or flower hue) come from the number of pangenes for that trait. Following Galton, de Vries plotted bell-shaped distribution curves for such variations, each around a stable mean. In contrast, he attributed species-causing variations (or mutations) to alterations in pangenes during reproduction. As they multiplied and spread through future generations, these altered pangenes introduced new traits into a population, de Vries believed, with each new trait establishing a new statistical mean and bell-shaped distribution pattern. If the new trait differs enough from the old one, then a new variety or species could result. Extensive breeding experiments generated numbers that seemed to fit the projected frequency patterns. A steady stream of publications followed, as de Vries documented the case for hard heredity and evolution by mutation. International recognition came to him and his small country, with de Vries clearly basking in it.[6]

Intense and humorless, de Vries drove himself and ignored his students. He wanted to understand what caused mutants, and labored to isolate their pangenes by crossbreeding them with normal plants of the same type. The answers lay in statistics, he realized, and among the relationships he explored was one suggested by Belgian mathematician Adolphe Quetelet. Random draws of two balls from an urn containing an equal (but effectively infinite) number of blue and red balls produces three chances in four of choosing at least one blue ball (or a 3:1 ratio of one or two blue balls versus two red balls). Quetelet had noted that this pattern should apply to biological phenomena involving a random selection of two opposite coded traits—which for de Vries included the inheritance of two opposing pangenes, such as one for a blue flower and one for a red one. Assuming one of the opposing pangenes dominates in manifesting its trait where both are present (which he may not have independently recognized), de Vries postulated that the results of crossbreeding hybrid plants (all containing one each of two opposing pangenes) should generate the 3:1 ratio of manifested traits in the next generation, such as three with blue flowers and one with red ones (where blue is the dominant trait). When de Vries found just such a statistical relationship in his breeding experiments, he saw it as further evidence for hard heredity and against blended inheritance. At some point in the discovery process, de Vries recognized that his findings conformed with those published thirty-four years earlier by Moravian botanist Gregor Mendel but largely forgotten thereafter. By 1900, when de Vries submitted his own findings for publication, he had boldly appropriated Mendel's reasoning without ac-knowledgment. "Modesty is a virtue, yet one gets further without it," de Vries later observed in a comment (made in a different context) that fit his actions in this episode. "I found it very difficult to find a middle way between German im-modesty and Dutch modesty."[7]

Unbeknownst to de Vries, German botanist Carl Correns had also uncovered the 3:1 ratio while crossbreeding hybrid plants during the late 1890s. He, too, had drawn on Mendel's long-neglected article to refine his thinking. When de Vries published his findings in 1900 without citing Mendel, Correns called him on it. Both botanists claimed to have independently discovered what Mendel already knew—and a third soon

joined this unseemly priority dispute—but Mendel understood and expressed the matter better than any of them, so
that his priority ultimately prevailed even though he had died
sixteen years before the rediscovery of his work. In 1866, when
originally published, Mendel's findings held little meaning for
biologists predisposed to think in qualitative terms of blended
inheritance and continuous variations. Those same findings
held profound significance in 1900, after Galton, Weismann,
and de Vries had accustomed biologists to think in statistical
terms about hard heredity and discontinuous variations. As a
result, Mendelism became one of the most significant biological discoveries of the twentieth century—more than three
decades after its initial announcement.

———

From 1856 to 1863, while serving as a science teacher supplied to a local technical school by the Roman Catholic
monastery near the Moravian capital of Brünn (now Brno),
Mendel had conducted the most extensive plant-breeding experiments hitherto reported by anyone. They involved growing, crossbreeding, observing, sorting, and counting nearly
thirty thousand pea plants of various carefully selected varieties. At the time, Moravia was a region of the Austro-
Hungarian Empire, then a protomodern police state with
quasi-medieval remnants of ecclesiastical privilege. Born of
peasant stock, Mendel received as good an education as anyone in his situation should have expected, including two years
at a regional university. With limited resources, however, his
most promising avenue of advancement became the established church.

In 1843, carrying a recommendation from his physics professor, Mendel obtained admission to the Augustinians'
wealthy and scholarly monastery near Brünn, where he remained for the rest of his life; he eventually became its abbot.
Enriched by its feudal land holdings, the monastery served as

a regional center for learning and scientific study, with many of its members teaching at local schools and some of them leaving to become university professors elsewhere. By all accounts, Mendel was an inspiring teacher, trustworthy administrator, and faithful monk. He certainly was a meticulous and persistent researcher. Nevertheless, even he could scarcely believe his success in life. Writing about himself in the third person, Mendel later observed, "When he looked back on his own past, as a peasant lad in Heitzeindorf, who had had so hard a struggle to achieve a high-school education, often ailing and always poor...he cannot but have been amazed to find himself at forty-six a mitred abbot."[8] His principal vices (if they can be called that) were fine food and good cigars, both of which as abbot he consumed in prodigious quantities.

Historians still cannot agree on what hypothesis Mendel was trying to test with his massive pea-plant-breeding experiments, but they were so carefully constructed that they must have had a clear purpose. Most likely, he wanted to see if new species arose from hybrids—an idea that dated to the pious eighteenth-century Swedish naturalist Carolus Linnaeus and, in the 1850s, still represented a modified-creationist alternative to the theory of evolution. In his published article, Mendel described his objective as simply being "to observe" the changes in "the hybrids and their progeny" that flow from crossbreeding pure varieties differing in only one or a few distinct traits. He knew from the literature and experience that, in such cases, common traits pass unchanged to the hybrids and their progeny, but that differing traits can take one form in the hybrid and other forms in the hybrid's progeny. Well-trained in mathematics, Mendel set about studying the process quantitatively by crossbreeding thousands of pea plants drawn from a few carefully selected varieties that consistently display one distinct variant of one or more of seven different paired traits. The paired traits were seed shape (round or angular), interior seed color (yellowish or intense green), seed-skin color (white or gray-brown), pod shape (smoothly arched or constricted), pod color (yellow or greenish), flower position (axillary or terminal), and plant height (tall or short).[9] Thus, for example, he crossbred tall varieties with short varieties and compared their heights with those of the hybrids and their progeny. Neglected for more than three decades, his results have since become famous.

In the hybrids produced during the first season of Mendel's experiments, for each pair of traits only one trait reappeared. Mendel called this the "dominating" trait. All the hybrids between tall and short varieties grew tall, for example, and all the hybrids between axillary- and terminal-flowered varieties produced plants with axillary flowers. There was no blending in this initial generation—no hybrids of medium height or bearing flowers both along the stem and bunched at its tip. This finding alone could have called into question the doctrine of blending inheritance, but even Mendel did not go that far. At the time, peas were seen as a somewhat special case in the way that traits pass through the generations, which was why Mendel chose them for his experiments. Many other hybrids appear to display a blend (or mixture) of parental traits, and it took more observations before biologists began seeing heredity as hard. Only then did they recognize hybrid peas as offering an unusually clear view of inheritance.

Mendel did not stop after the first season, of course. Over the ensuing years, he bred his hybrids with others of their types and then did the same with their progeny. In the second season, for each of the hybrid types, both of the original paired traits reappeared in their original forms among the progeny. In every case, approximately three-fourths of the progeny displayed the dominant trait and one-fourth displayed its opposite pair, which Mendel called the "recessive" trait. From breeding the tall hybrids, Mendel got 787 tall plants and 277 short plants, for example, and from the axillary flowered hybrids, there came 651 with flowers along the stem and 207 with flowers at stem tips. In all cases, the ratio between dominant and recessive traits approached 3:1, with no blending. In subsequent generations, breeding together the progeny with recessive traits produced only plants with the recessive trait, while breeding together the progeny with dominant traits produced one-third that consistently displayed the dominant trait and two-thirds that split according to the 3:1 ratio of dominant to recessive. Experiments with multiple-character pairs suggested that each pair segregated independently.

Although Mendel did not anticipate such developments in his articles, by 1900 some biologists could begin to see how his results fit into their emerging understanding of hard hereditary information carried on chromosomes in discrete units they came to call "genes." If each parent plant carries two genes for each trait and passes along only one of them (picked at random) in reproduction, then Mendel's results made perfect sense. Breeding together pea plants having two genes for tall with ones having two genes for short would produce hybrids with one gene for tall and one gene for short, for example. If tall is dominant, then all the hybrids would grow tall while carrying a masked gene for short. This could explain Mendel's findings for the first season. When the hybrids are bred, however, there is an equal chance of passing one gene for either paired trait from each of the parents—creating the type of statistical relationship described by Quetelet. As with blue and red balls drawn from an urn, the result is a one-in-four chance of two genes of either trait and a two-in-four chance of one gene of both traits. If one of the traits is dominant, then the 3:1 ratio of dominant to recessive appears in the second season. In the third season, breeding together progeny carrying two identical genes of the same trait produces offspring displaying that trait consistently, while breeding together progeny carrying two differing genes for the trait produces offspring displaying the 3:1 ratio.

Whether Mendel understood his findings this way in 1866 made little difference by 1900, when his work was rediscovered. By then, this was gradually becoming the logical interpretation of his data. Indeed, it was a turn-of-the-century Mendelian, the Danish botanist Wilhelm Johannsen, who gave the name "gene" to the hypothetical unit of hereditary information—a contraction of de Vries's term "pangenes." Mendel had referred to them interchangeably as "factors," "characters," and "cells." He probably envisioned them as immaterial essences—after all, he was an Augustinian monk, and that is how Augustine probably would have viewed them.

———

During the first decade of the twentieth century, various European and American biologists began taking up Mendel's work and exploring its implications for evolutionary theory—something Mendel never did. Of course, all these scientists brought their own conceptions of the problem to the effort. In the scientific mind, there is no such thing as a clean slate; indeed, often quite a bit of erasing is required before anyone can write something new on the board. This certainly applied to Pearson, Weldon, and the other early biometricians who championed the role of continuous, incremental variations in the evolutionary process. They linked Mendelian concepts of discontinuous heredity (based in opposing paired characteristics) with saltationist theories of discontinuous evolution, and expanded their attacks on the latter to include the former. None of them viewed Mendelism as a significant development in science; Weldon went so far as to denounce it as a fraud.

Mutation theorists proved more responsive than biometricians to Mendel's so-called "laws of heredity." Promptly after the rediscovery of Mendelism in 1900, they started testing the

range of its application in biology. Although de Vries soon fell
out from this crowd when he concluded that his species-causing
mutations did not follow Mendel's laws, Bateson, Johannsen, and
many other proponents of theories of discontinuous evolution
more than took up the slack. They began calling themselves "geneticists" and their field "genetics." For them, Mendel's laws suggested a means of propagating gene-based variations without
regression, swamping, or acquired characteristics. If a mutant
genetic trait was dominant, then under Mendel's laws it could
sweep through a population rapidly without any loss through
blending. Even a recessive genetic mutation could survive unal-
tered in a population. Natural selection at most serves to weed
out grossly unfit mutations, these early geneticists typically believed, and thus it plays only a secondary role in the evolutionary process. Mutations generate self-propagating variations,
they maintained, and evolution naturally results.

Despite their enthusiasm for Mendelism, Bateson and Johannsen refused to see genes as material particles on chromosomes. Instead, they speculated that genes might exist as
immaterial waves or energy states distributed throughout the
cell or organism. It fell to two other early Mendelians,
Theodor Boveri and Walter Sutton, to point out that hypothetical Mendelian factors behave like observed chromosomes, in that, for any given organism, they come in mated
pairs, with one mate derived from each parent. Perhaps, they
noted independently in 1902, chromosomes carry Mendel's
factors. Such thinking remained largely theoretical until a
team of Columbia University researchers led by the skeptical
Kentuckian Thomas Hunt Morgan sealed the bond between
Mendel's laws of heredity and material chromosomes during
the early 1910s. In so doing, Morgan's team laid the groundwork for the modern synthesis of genetics and Darwinism
that has dominated biological thought ever since.

Morgan owed his triumph, which duly netted him the first

Thomas Hunt Morgan in his laboratory with bottles containing fruit-fly specimens, 1922.

Nobel Prize awarded to an American biologist, to a roomful of fruit flies. He deserves the credit for choosing them as his research subjects, but they showed him something he never expected to see: genes stretched along chromosomes "like beads on a string."

When Morgan began this research in 1908 or 1909, he was already a highly respected biologist. Born in 1866 into an aristocratic Southern family devastated by the outcome of the Civil War, and trained in the naturalist tradition, by 1900 Morgan had grown to regard the experimental method of quantitative, laboratory-based research as the only valid means to investigate the chemical and physical processes that constitute life. There are no metaphysical purposes or vital spirits that differentiate life from nonlife, he maintained, and thus proven methods from the physical sciences should apply in biological research. "Hard as it is to image, inconceivably hard it may appear to many, that there is no direct relation between the origin of useful variations and the ends they come to serve, yet the modern zoologist takes his stand as a man of science on this ground," Morgan wrote in 1909. "He may admit in secret to his father confessor, the metaphysician, that his poor intellect staggers under such a supposition, but he bravely carries forward his work of investigating along the only lines that he has found fruitful."[10] Sentiment and Southern manners softened its outward appearance, but reductionism shaped Morgan's scientific worldview. There was no place for souls or supernatural beings in his work, though he played Santa Claus for his children every Christmas.

———

Drawn to the big questions, Morgan naturally wanted to know how evolution operated. He dismissed Darwinism for lacking a credible explanation for heredity and criticized its most prominent living proponent, August Weismann, for engaging in flights of fancy. "Invisible germs whose sole functions are those which Weismann's imagination bestows on them, are brought forward as though they could supply the deficiencies of Darwin's theory," Morgan complained in his 1903 book, *Evolution and Adaptation*. Lamarckism fared no better in his estimation because its many proponents failed to demonstrate that acquired characteristics are inheritable. "Despite the large number of cases that they have collected," Morgan noted in his 1903 book, "the proof that such inheritance is possible is not forthcoming. Why not then spend a small part of the energy, that has been used to expound the theory, in demonstrating that such a thing is really possible?"[11]

Mutation theory, in Morgan's mind, was the most credible explanation for the origin of species. He admired how de Vries developed his theory through large-scale flower-breeding experiments. Morgan sought to test and extend it using other species. He settled on the fruit fly, *Drosophila melanogaster,* because it is quick, easy, and cheap to breed in great numbers. "It's wonderful material," Morgan boasted in 1910. "They breed all year round and give a new generation every twelve days."[12] Within the first six years, by which time Morgan had made his most profound discoveries, his research team had watched more generations of them go by than Mendel or de Vries could have observed with peas or primroses in two centuries.

Morgan began his fruit-fly research skeptical of Mendel's laws of heredity and their supposed basis in genes on paired chromosomes. Both in nature and the laboratory, many inherited traits seem to take on intermediate (rather than pure) forms in successive generations, he noted, and the classic case of a distinct paired trait, sex, follows a 1:1 (rather than 3:1) ratio. Further, Morgan added, if each chromosome carries multiple genes, as they must given the small number of chromosomes and large number of traits, then breeders should report more instances than they do of linked traits inherited together on a shared chromosome. The very notion of dominant and recessive genes carried on chromosomes lacked sufficient experimental evidence to convince Morgan, and he had not intended to test it. Instead, he was looking for species-causing mutations, such as de Vries had found among primroses, and wanted to see if they bred true: a very un-Mendelian objective.

It took Morgan's team about a year of breeding flies to spot one with a recognizable mutation—a white-eyed male in a roomful of red-eyed flies. It did not rise to the level of a species-causing mutation, but at least it was discrete and discontinuous. He bred the mutant with a normal, red-eyed female, then inbred their offspring. All 1,237 flies in the first generation had red eyes, but white-eyed flies reappeared in the next—approximately one for every three with red eyes. Morgan had found Mendel's ratio with a recessive mutation in fruit flies. Further, all the white-eyed flies were male—the traits for sex and eye color were linked as if carried on the same chromosome pair.

Breeding fruit flies was messy business. Over time, Morgan accumulated a corps of student assistants to help him. They worked in a fifteen-by-twenty-foot sixth-floor lab, nicknamed "the fly room," on the Columbia University campus. Half-pint milk bottles filled with flies covered desks and lined shelves in the lab. The stench of rotten bananas (to feed the flies) and ether (to anesthetize them for study) filled the air, along with swarms of escaped subjects. Morgan and his students worked long hours in the fly room, but a generation of geneticists received their graduate or postgraduate training there. Morgan created a remarkably fluid research environment, with students free to share their insights and pursue promising leads. Three of them stood out—Alfred Sturtevant, Calvin Bridges, and Hermann Muller—and Morgan co-authored with them the epoch-making 1915 book *The Mechanism of Mendelian Inheritance,* which reported their findings to the scientific world. Finding the white-eyed fruit fly in 1910 was followed soon by the discovery of two more sex-linked mutations, yellow body color and miniature wings, both of which reproduced in Mendelian ratios. By then, Mor-

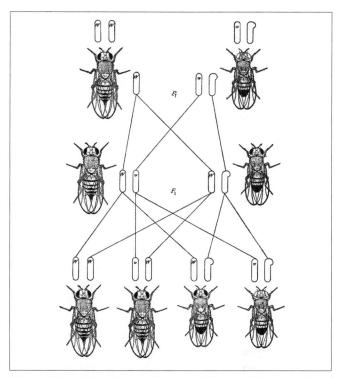

Thomas Hunt Morgan's 1915 diagram of the Mendelian genetics of eye color in fruit flies, with the mother's egg containing an X chromosome carrying a gene for red eyes and the father's sperm containing either an X chromosome carrying a mutant gene for white eyes or a Y chromosome with no gene for eye color.

gan had become a believer in both Mendelism and the chromosome theory of heredity.

Once Morgan and his students had mastered the enterprise, discoveries and insights poured out of the fly room. Between 1911 and 1915, they identified dozens of mutations.

Through breeding experiments, they determined that all the mutant traits fell into one of four linkage groups corresponding to the number of chromosome pairs in fruit-fly cells. The genes for each linked trait typically reside on the same chromosome pair, Morgan concluded, although not always. Eye color normally linked with sex, for example, but exceptions occurred, such as in the rare cases of white-eyed females. Experience showed that some links were stronger (or exhibited fewer exceptions) while other links were weaker (or exhibited more exceptions).

To account for these experimental results, Morgan turned to the work of Belgian cytologist Franz Janssens. In 1909, Janssens observed that, during the process of meiosis, chromosomes occasionally cross over each other, break apart at corresponding points, and reconnect with an exchange of parts. Such recombinations would disrupt gene linkages, Morgan reasoned, and the farther apart two genes lay on a chromosome, the more likely that a break could occur between them. Variations in the strength of different gene linkages suggested to Morgan that genes must lie along the length of chromosomes, like beads on strings that occasionally break during meiosis. Sturtevant noted that this phenomenon, coupled with data on the relative strengths of various gene linkages, should allow researchers to map the location of specific genes on various chromosomes. Without ever seeing a gene, by 1915 Morgan and his students had used their study of mutant flies to establish the existence of genes, map their locations on chromosomes, and elucidate the basic elements of classical genetics.

The Mechanism of Mendelian Inheritance effectively completed a revolution in scientific thought that placed genes at the center of biologists' conception of heredity, pushing environmental and developmental factors to the periphery (just as

Francis Galton had wanted). In time, these tiny material particles would become the foundation blocks for a new synthesis in evolutionary thought.

Morgan never fully passed over from the wilderness of mutation theory to the promised land in which classical genetics and selection theory combine to form the modern neo-Darwinian synthesis. An avowed empiricist, Morgan only trusted theories he could confirm by laboratory experiment—and natural selection proved difficult to test in that way, especially with fruit flies maintained in artificial conditions conducive to their survival. Experiments showed him that mutations were smaller than he had originally thought, and that they were preserved and propagated in a Mendelian manner, but his basic view did not change. He payed little attention to other sources of genetic variation, such as gene flows across species. New types of organisms evolve over time through the accumulation of random, inborn mutations, Morgan maintained, with natural selection serving mainly to nip detrimental mutations in the bud. He minimized the role of adaptation, giving scant credence to the idea (central to the modern synthesis) that selection could act on existing genetic variation within a population in adapting organisms to their environment. With his original view reinforced by his *Drosophila* research, Morgan relied on mutations to do virtually the entire work of evolution. Even the pace of evolution, which most Darwinian naturalists explained in terms of natural selection, succumbed to a mutationist interpretation when Muller showed in 1927 that radiation exposure accelerates the rate of mutation in fruit flies. Evolution could thus speed up and slow down in response to fluctuation in the level of natural radiation independent of natural selection.

Morgan's modified Mendelian version of the mutation theory remains a key part of orthodox evolutionary thought, but far from the whole story. Some of his own students would broaden the account considerably by integrating insights from field biology and paleontology. Yet without Morgan's exposition of classical genetics, that larger story would have remained a mystery.

CHAPTER 8

APPLIED HUMAN EVOLUTION

E ven his friends and admirers viewed Francis Galton as an eccentric genius. Perhaps it was hereditary. After all, the grandfather he shared with his cousin Charles Darwin was the brilliant but idiosyncratic physician Erasmus Darwin, who died before either Francis or Charles was born. Along with maintaining a lucrative medical practice in which he stressed the power of mind over illness and liberally administered opium to make his patients feel better, Erasmus Darwin wrote chic countercultural poems and medical treatises laced with his radical political, religious, and scientific views, including wonderfully provocative speculations about organic evolution. Befitting an author who sired at least fourteen children in or out of wedlock, he also extolled the joys of sex. The elder Darwin enjoyed eating, too, and he grew so stout that he had a semicircle cut from his dining table for his large stomach so he could remain within arm's length of his food. In spirit if not in politics or diet, Galton fell closer to Erasmus Darwin's side of the family tree than did his more restrained cousin Charles.

Something of a child prodigy, Galton remained inventive throughout his long life. He learned to read at age two, but partied more than he studied during his college years at Cambridge—and graduated without honors. Then came his enormous inheritance and African adventures. In his thirties, Galton married and settled into the life of a wealthy, well-connected gentleman scholar, dividing his time between his elegant London townhouse and excursions on the European continent. Beyond his foundational contributions to genetics and statistics, he conducted significant research in cartography, geography, meteorology, psychology, and sociology. Except for his sustained work in eugenics, Galton typically offered brilliant insights in a field (such as recognizing the phenomenon of anticyclones in meteorology) but showed little follow-through, leading some to dismiss him as a dilettante. He also invented a host of clever gadgets, from underwater reading glasses and high-pitched whistles for testing hearing to a sand-glass speedometer for bicyclists and a self-tipping top hat. More lasting, he pioneered the use of fingerprinting for personal identification.

At heart Galton was a snob, and his pedantic manner kept him at a distance from Charles Darwin's inner circle. A telling example of Galton's character occurred when, as a section president of the British Association for the Advancement of Science, it fell upon him to introduce American journalist Henry Stanley for a public lecture at the association's 1872 annual meeting. Stanley had just returned from his celebrated African expedition to relieve the revered Scottish missionary-explorer David Livingstone. In the glare of publicity surrounding that feat, a rumor (later confirmed) held that Stanley was the illegitimate son of a Welsh farmer. Galton's introduction, rather than praising Stanley, expressed the hope that he would clarify mysteries about his pedigree. After Stanley finished, Galton criticized him for telling adventure tales without scientific substance. The fact that Livingstone, whom Stanley hailed in his lecture, was an evangelical hero surely helped to prompt Galton's sneers. Galton was contemptuous of Christianity, and once tried to disprove the efficacy of prayer by demonstrating statistically that prayed-for people (such as monarchs or the sick in Christian households) do not live longer or recover more quickly than others. For Galton, Darwinism both justified elitism and undermined the Christian doctrines of divine

Francis Galton, from a
portrait made shortly
after the publication of
Hereditary Genius in 1869.

creation and original sin. Eugenics emanated from this potent mixture of brilliance and bias.

SOWING EUGENIC SEEDS

Galton coined the word "eugenics" in his 1883 book, *Inquiries
into Human Faculty and Its Development*, which appeared nearly
two decades after he began trying to make a science out of
human breeding. He summarized his mature conception in
the book's introduction. "My general object has been to take
note of the varied hereditary faculties of different men," Galton explained, "to learn how far history may have shown the
practicability of supplanting inefficient human stock by better strains, and to consider whether it might not be our duty
to do so by such efforts as may be reasonable, thus exerting

ourselves to further the ends of evolution more rapidly and with less distress than if events were left to their own course."[1] As a Darwinist, he believed that humanity had risen to its present level through an evolutionary process driven by the natural selection of persons with beneficial inborn traits. Galton's scheme involved identifying the traits that advance humanity and then artificially selecting persons with them to dominate in reproducing the next generation. With the advent of eugenics, he promised, "what Nature does blindly, slowly, and ruthlessly, man may do providently, quickly, and kindly."[2] For Galton and his followers, this became the utopian vision of applied human evolution; for their critics and many victims, it became a dystopia.

Galton derived the word "eugenics" from the Greek for "well-born," signaling his fascination with the sources of natural ability rather than the causes of human disability. His earliest publications on the topic, an 1865 magazine article titled "Hereditary Talent and Character" and the 1869 book *Hereditary Genius,* systematically presented the pedigrees of eminent men, whose names Galton culled from biographical reference books and lists of high achievers, leading him to conclude that genius runs in families. Galton later pioneered the comparative study of twins raised separately. "Everywhere is the enormous power of hereditary influence forced on our attention," he wrote, "proving the vast preponderating effects of nature over nurture."[3] On the basis of these findings, Galton proposed that society should encourage men and women of hereditary fitness to marry each other and to bear many children—propositions that became known as "positive" eugenics. "What an extraordinary effect might be produced on our race, if its object was to unite in marriage those who possessed the finest and most suitable natures, mental, moral, and physical!" he exclaimed.[4] Reflecting his belief that physical beauty reflected eugenic fitness, Galton collected data for a beauty map of the British Isles. "I found London to rank highest for beauty; Aberdeen lowest," the English scholar observed.[5]

Although Galton urged from the outset that society also discourage "weakly and incapable" persons from breeding, and later recommended their compulsory segregation into single-sex institutions, he never focused on so-called "negative" eugenics as much as did his followers in America and Germany.[6] Combining his belief in hard heredity with his faith in the cumulative impact of marginally raising the reproductive rate for the fit and marginally lowering it for the unfit, Galton affirmed "that the improvement of the breed of mankind is no insuperable difficulty."[7] Indeed, he predicted, "If a twentieth part of the cost and pains were spent in measures for the improvement of the human race that is spent on the improvement of the breed of horses and cattle, what a galaxy of genius might we not create!"[8] For Galton, this represented the promise of applied evolution and the highest mission of a scientific society. "I take Eugenics very seriously, feeling that its principles ought to become one of the dominant motives in a civilized nation, much as if they were one of its religious tenets," he wrote in the concluding passage of his autobiography. In the past, Galton explained, human evolution proceeded fitfully and with great individual pain through natural selection, but with the developed mind of modern man, "I conceive it to fall well within his province to replace Natural Selection by other processes that are more merciful and not less effective."[9]

For Galton, eugenics would operate solely within a race, and not across racial groups. Like Darwin and Haeckel, he believed in a strict hierarchy of racial types, with some subset of northern Europeans at the apex of the evolutionary pyramid. All three of them assumed that where different races came into extended contact, superior ones would inevitably supplant inferior ones through natural selection. Galton never suggested that artificial selection either should hasten this process or could alter its outcome. "There exists a sentiment, for the most part quite unreasonable, against the gradual extinction of an inferior race," Galton explained. "It rests on some confusion between the race and the individual, as if the destruction of a race was equivalent to the destruction of a large number of men. It is nothing of the kind when the process of extinction works silently and slowly through" natural selection.[10] These comments reflect the mainstream tenets of late-nineteenth-century scientific race theory. Darwin made precisely the same arguments in *Descent of Man.*[11] Only in the twentieth century would some of Haeckel's followers attempt to assist the process through artificial mass selection via the Nazi death camps.

Where Galton concentrated his research on bloodlines of distinction, pioneering the genealogy of degeneracy fell to a New York social reformer, Richard Dugdale. He became interested in the issue when, during an 1874 inspection of conditions at a jail in rural New York State, he learned that six of the prisoners were related. Struck by this observation, Dugdale launched an investigation into the lineage of their family, which he called "the Jukes," in an effort to uncover the causes of crime. Using prison records, relief rolls, and court files, he traced the Jukes family tree through five generations. Dugdale found that more than half of the 709 people related to this family by blood or marriage were criminals, prostitutes, or destitute.[12] The publication of Dugdale's findings in 1877 caused a sensation and spawned a small genre of similar studies, all reaching essentially the same conclusion: Degeneracy, like genius, runs in families.

Galton's publications also stirred considerable interest. *Hereditary Genius* appeared just as Darwin was finishing *Descent of Man,* and he included his younger cousin's findings within that book. "We now know through the admirable labors of Mr. Galton that genius," Darwin wove into his text, "tends to be inherited; and, on the other hand, it is too certain that insanity and deteriorated mental powers likewise run in the same families."[13] Like the work of many eccentric geniuses, however, Galton's eugenics was somewhat ahead of its time. For his part, Dugdale simply never anticipated how eugenicists would later use his findings because (unlike Galton) he was not one of them. Neither Galton's 1869 *Hereditary Genius* nor Dugdale's 1877 *The Jukes: A Story in Crime, Pauperism, Disease, and Heredity* provoked a rush to administer eugenic remedies for perceived social diseases. The science of the day did not yet justify them.

Subscribing to the then-prevailing blending view of heredity, Galton saw the challenge not so much as pulling inferior families up to the norm (because that should happen naturally through crossbreeding) but rather in keeping superior families from falling back toward it. Thus, Galton stressed positive eugenics. Discouraging reproduction by the unfit, which later became the central thrust of the eugenics movement, did not initially appear as critical as did encouraging the fit to intermarry and breed. Although many early eugenicists worried openly about the allegedly low birthrate for families in elite society (and some even spoke excitedly about "race suicide" as its consequence), in reality members of the Victorian social elite tended to marry within their class and have large families anyway. Perhaps the issue became exaggerated for Galton as his own marriage proved barren—a heavy cross to bear for the father of eugenics.

Another scientific factor limiting the appeal of eugenics was the widespread acceptance (though not by Galton) of Lamarckian notions of soft heredity. It was not that Lamarckians never embraced eugenic remedies—many of them later did; it was simply that a belief in the inheritance of acquired characteristics undercut the urgency of such prescriptions. This was apparent in Dugdale's recommendation of how to deal with the Jukes. "Environment tends to produce habits which may become hereditary, especially so in pauperism and licentiousness," he explained in patent Lamarckian terms. "The correction is change of the environment." In particular, Dugdale argued, the way to break a chain of hereditary degeneracy is to remove the children of social misfits from their parents and raise them in a healthy environment. Eugenic restrictions on reproduction were thus unnecessary because, as Dugdale asserted, "Where the environment changes in youth, the characteristics of heredity may be measurably altered."[14]

By the beginning of the twentieth century, the scientific theories holding eugenics back began to crumble. Galton, Weismann, de Vries, and others pressed the case for hard heredity. Then, in 1900, the rediscovery of Mendel's laws suggested that parental and ancestral traits reappear in children and more remote descendants without blending. Thus, if there are superior and inferior hereditary traits, and if their impact on succeeding generations is unalterable by environmental influences or blending (as suggested by Mendelism), then the scientific case for eugenics could appear compelling—especially if the genes (or germ plasm) carrying those traits are viewed as operating like simple Mendelian factors. "More children from the fit, less from the unfit," became the motto of a new generation of eugenicists.[15] Of course, the triumph of eugenics was built on a history of greater public acceptance of both hereditarianism and of a competitive struggle for existence. It only took a slight twist of reasoning to transpose accepting the natural selection of the fit into encouraging the intentional elimination of the unfit. As applied to human society and behavior, all such survival-of-the-fittest thinking became known to its critics as "Social Darwinism," even though its origins predated Darwin. Because of the energy it drew from the scientific authority of Darwinism, however, the development of Social Darwinism fed into the larger history of evolutionary thought.

THE SPROUTING OF SOCIAL DARWINISM

To value competition fit the spirit of the age; its roots extended long before the publication of *Origin of Species.* Indeed, Darwinism represented simply one among many logical developments of an increasingly pervasive Western mindset that accepted competition among people or groups of people as socially beneficial. During the late 1700s, Adam Smith argued that economic progress depended on individual competition. His faith in the natural harmony of human interactions gave him hope that all people would benefit from *laissez-faire* capitalism. Embracing *laissez-faire,* Thomas Malthus soon observed that some individuals must gain and others lose in any social competition due to limited resources. Referring to the process as a "struggle for existence" (at least in the context of primitive human societies), Malthus wrote of the "goad of necessity" bringing out the best in people.[16] As early as 1851, in his breakthrough book *Social Statics,* Herbert Spencer began sketching out his concept that a form of natural selection, which he termed "survival of the fittest," worked hand-in-hand with an essentially Lamarckian type of evolution to generate human progress over time. Since it continually culled the unfit, Spencer saw selection as maintaining human quality.

With *Origin of Species,* Darwin pushed this line of reasoning a critical step further by presenting competition as producing fitter varieties, races, and, ultimately, species. Spencer and many other Victorian social scientists quickly accepted the key Darwinian insight: Regardless of the source of variations (whether chance, acquired characteristics, internal factors, or even God), all aspects of human nature and behavior, like everything else in the biological world, originate and evolve through the selection of individuals who display particular traits. Nothing was exempt, not even altruistic behavior or belief in the divine, both of which Darwin in *Descent of Man* attempted to explain in terms of their survival values for the individual or group. In its broadest sense, this was Social Darwinism, and its influence percolated throughout the social sciences and popular cultures in Europe and America.

Darwin was far from alone in seeking evolutionary explanations for human nature. His British disciple George Romanes joined in trying to find animal origins for human mental traits. Other social scientists in Europe and the United States did so, as well. Looking at the issue from the opposite end, Italian criminologist Cesare Lombroso, another self-proclaimed disciple of Darwin, saw much antisocial behavior as a throwback to humankind's savage ancestry. Born criminals, the criminally insane, and epileptics simply had failed to develop to the evolutionary level of their race, he explained, and are left behind as so-called "moral imbeciles."[17] Lombroso's theories attracted a wide following during the late 1800s, as European and American social scientists struggled to account for the seeming explosion of crime, mental illness, mental retardation, and poverty afflicting modern society. Civilization was evolving so rapidly from an agrarian to an industrial lifestyle, they reasoned, that an increasing number of people were hereditarily unable to keep up.

Industrialization and urbanization transformed western Europe and the United States during the late nineteenth century. Manufacturing boomed and people crowded into the cities. Social Darwinism sanctioned cutthroat competition in business and disparaged government efforts to help the needy. " 'They have no work,' you say," Spencer mocked those pleading on behalf of London's growing underclass. "Say rather that they either refuse work or quickly turn themselves out of it. They are simply good-for-nothings, who in one way or other live on the good-for-somethings."[18] The advances of civilized life had allowed the unfit to survive and multiply, he claimed, so that they threatened to swamp those responsible for creating modern civilization. To rectify the situation, Spencer urged government to stop interfering in economic and social affairs. Regulation slowed progress, he claimed, while public-health and welfare programs harmed people in the long run by preserving and multiplying the unfit.

Social Darwinism had its advocates throughout the Western world. Espousing the motto "root, hog, or die," for example, American political economist William Graham Sumner characterized competition as "the iron spur which has driven the race on to all which it has ever achieved." Neither public welfare nor private charity should restrain the natural struggle for existence, he stressed in an 1881 essay: "The law of the survival of the fittest was not made by man and cannot be abrogated by man. We can only, by interfering with it, produce the survival of the unfittest."[19] Darwin's translator, Clémence Royer, made similar arguments in her long preface to the French edition of *Origin of Species* and in her 1870 book, *The Origin of Man and Societies.*

Social Darwinism influenced popular culture, as well. Gilded Age capitalists such as John D. Rockefeller and James J. Hill publicly justified their monopolistic business practices in survival-of-the-fittest terms. Opponents of public-health and welfare programs drew on Social Darwinist thinking to claim that personal freedom demanded nothing less than an end to social legislation—leading U.S. Supreme Court Justice Oliver Wendell Holmes to complain bitterly, "The Fourteenth Amendment [of the federal Constitution] does not enact Mr. Herbert Spencer's *Social Statics.*" Holmes wrote these words in a dissenting opinion, however. The court's majority in that landmark case, *Lockner v. New York,* applied Social Darwinian reasoning to strike down a state worker-protection statute.[20] Countless writers, musicians, and other artists played with Social Darwinian themes in their compositions: Theodore Dreiser, Edith Wharton, and Richard Wagner offer familiar examples. A classic literary scenario involved four shipwrecked persons aboard a lifeboat in Stephen Crane's highly popular 1897 story "The Open Boat," in which the writer explored the character traits best fitting humans for survival in the clutches of nature's indifference, with brute strength losing out in the end.

While not all evolutionists accepted Social Darwinism, many Social Darwinists invoked evolutionary science to support their economic and social views. Historians rightly question the extent to which a strictly Darwinian theory of evolution informed this social discourse, but surely evolutionary naturalism and selectionism underlay much of it.

———

Although Darwin stressed the formative role of competition among individuals in his theory of evolution, this was but one field of battle for many late-nineteenth-century Social Darwinians. Some of them, such as Haeckel in Germany and Georges Vacher de Lapouge in France, saw competition among races or nations as more crucial for human evolution than any residual forms of interpersonal competition. Whereas Spencer's emphasis on individual competition tended to minimize the state's role in society, Haeckel's stress on racial and national competition tended to maximize it. Social Darwinism had many faces. Just when some Social Darwinists called for less government interference in domestic affairs, others championed imperialism, colonialism, and militarism in foreign affairs. Both scientific racism and militant nationalism became hallmarks of Social Darwinism, and it made little practical difference whether their proponents believed in a Lamarckian or

Darwinian theory of evolution: Either could justify racism or nationalism for persons already so inclined.

Racism predated Social Darwinism, of course, but for many Social Darwinists, the theory of evolution seemed to support their sense of racial superiority. Many Lamarckians saw the various human races as representing different stages of linear biological development, with the taxonomic status of each reflected in its relative cultural attainments. Befitting his Lamarckian orientation, Spencer not only believed in a biologically based hierarchy of races, but thought that all individuals, as they matured, recapitulated the evolutionary history of their race. "During early years every civilized man passes through that phase of character exhibited by the barbarous race from which he is descended," he explained. "Hence the tendencies to cruelty, to thieving, to lying, so general among children."[21] Haeckel, too, with his Lamarckian perspective, subdivided humanity into an intricate hierarchy of evolving races and species, with "the Germanic race, in North-western Europe and in North America," on top.[22] Social Darwinists in the United States, such as Sumner and the noted Lamarckian geologist Joseph LeConte, drew on such views to justify the continued political subjugation of blacks in the post–Civil War South. "The Negro race is still in childhood," the Georgia-born LeConte opined in 1892; "it has not yet learned to walk alone in the paths of civilization."[23]

Despite Darwin's view of evolution as branching rather than linear, with nothing inherently progressive about it, most Darwinian scientists joined their Lamarckian counterparts in positing a single line of human development. Some white ethnic subgroup blossomed at the end of this long, solitary branch, they inevitably concluded, whose supposed superiority they typically attributed to the invigorating challenge of surviving in a cold climate. "Extinction follows chiefly from the competition of tribe with tribe, and race with race," Darwin wrote in *Descent of Man*. "When civilized nations come into contact with barbarians the struggle is short, except where a deadly climate gives its aid to the native race."[24] Lapouge was less sanguine than Darwin about which races would prevail in the ongoing struggle for existence. An anthropologist without standing in his own country but with an influential following in Germany and the United States, Lapouge compulsively calibrated racial hierarchies based on skull shape, but worried obsessively that inferior, round-headed races might overrun his superior, long-headed "Aryan" race. "Evolution takes place all around us," he explained in his 1899 book, *L'Aryen*. "It does not indefinitely lead toward the better, it leads toward nothing."[25] Such worries, carried to the extreme by a small corps of radical evolutionists, fed a racist variant of eugenics that advocated government policies of ethnic exclusion or elimination.

Although it made little sense from a biological perspective, some Social Darwinists called for militaristic competition among nations. Darwin, Spencer, and even Lapouge vehemently disagreed, fearing that war weakened a civilized society by killing its ablest young men. Nevertheless, Haeckel advocated a strong, unified Germany to dominate the world. "Nowhere in nature," he wrote in his popular 1868 *History of Creation*, "does that idyllic peace exist, of which poets sing; we find everywhere a struggle and a striving to annihilate neighbours and competitors." Thus, he stressed, "the whole history of nations... must therefore be explicable by means of *natural selection*. Passion and selfishness—conscious or unconscious—is everywhere the motive force of life."[26] This Social Darwinian vision of national progress fed German militarism leading up to World War One. During that bloody conflict, American evolutionary zoologist Vernon Kellogg, then on a failed peace mission to Europe, concluded that a "Neo-Darwinian struggle-for-existence" mindset propelled the intellectual elite of the

German officers' corps.[27] It was a profoundly disquieting discovery for Kellogg, and one that soon helped to launch a popular crusade against evolutionary biology in the United States.

Nationalistic competition, like racial competition, dovetailed with the eugenics movement, which gained momentum following the rediscovery of Mendel's laws in 1900. Eugenics quickly became the focal point of applied human evolution, and remained so at least until the 1930s. It took two complementary forms, positive eugenics (or "more children from the fit") and negative eugenics (or "less [children] from the unfit").[28] The former was typically voluntary; the latter became increasingly compulsory.

EUGENICS IN FULL BLOOM

In the early twentieth century, the newly energized eugenics movement built on the pioneering efforts of Galton and Dugdale. The Eugenics Education Society, championing negative eugenics in Britain, recruited an elderly Galton to serve as its honorary president. The Galton Laboratory for National Eugenics at University College, London, funded by Galton and directed by his protégé Karl Pearson, acted as the society's research arm. Galton's first cousin once removed, Charles Darwin's middle son Leonard, served as its president from 1911 to 1925. The leading American eugenics organization, the Eugenics Record Office at the Carnegie Institution's Cold Spring Harbor genetics laboratory, revised and distributed Dugdale's classic study of hereditary degeneracy in 1915.

The revised study of the Jukes family revealed much about the transformation in social-scientific thought brought about by the rise of Mendelism. Where Dugdale's original book categorized family members by their social behaviors, the revised edition did so by their supposed mental abilities. The new study, which included two thousand more family members (most of them living), concluded that "over half the Jukes were or are feebleminded."[29] This finding was significant because eugenicists typically viewed "feeblemindedness," which covered various levels of mental deficiency, as an inherited Mendelian trait. "If the Jukes family were of normal intelligence, a change of environment would have worked wonders," explained psychologist Henry H. Goddard of the prestigious Training School for Feeble-Minded Boys and Girls in Vineland, New Jersey. "But if they were feebleminded, then no amount of good environment could have made them anything else than feeble-minded."[30] Low intelligence breeds antisocial behavior, eugenicists assumed, and its propagation weakens the race. Accordingly, the revised study recommended permanent sexual segregation or sterilization for all the Jukes (where Dugdale had prescribed a change of environment for the younger Jukes). The 1915 study reflected the means and objectives of a mature eugenics movement.

The first task faced by eugenicists was to identify those who should not reproduce. Hereditary forms of mental defect and deficiency became a main target. Goddard focused his attention on those deemed mentally deficient, suggested a mental age of thirteen years as the lowest appropriate level for reproduction, and coined the term "moron" (from the Greek for "foolish") to identify adults whose mentality fell just below this minimum.[31] He imported the Binet-Simon intelligence test from France as a means to compute mental age or, as later refined, "I.Q."

Influenced by Lombroso's work in Italy, some eugenicists also targeted repeat criminals, prostitutes, and others who regularly manifested certain, supposedly hereditary, undesirable social behaviors. Such physical conditions as epilepsy, hereditary blindness, and assorted gross deformities were singled out as grounds for restriction, as well. Some enthusiasts urged casting the net far wider: journalist H. L. Mencken only half-jokingly proposed a program for the mass sterilization of sharecroppers in the American South.[32] Even a special investigatory committee of the American Neurological Association, whose 1936 report sharply criticized eugenic excesses in the United States, recommended sexual sterilization for certain disabilities, including some inherited forms of mental illness and retardation, "disabling degenerative diseases recognized to be hereditary," and epilepsy.[33]

Much of the historical analysis given to these matters focuses on compulsory state programs designed to stop breeding by persons classified as mentally defective or deficient. Goddard and Eugenics Record Office director Charles Davenport (a noted geneticist) advocated marriage restrictions, forced segregation during the reproductive years, and compulsory sterilization. Others offered more drastic remedies, including infanticide and euthanasia. In France, Lapouge darkly warned of future "copious exterminations of entire peoples" if the government failed to impose strict limits on human breeding.[34] After enough public-health officials, mental-health experts, physicians, and social reformers (many of them women) echoed these calls, lawmakers and government leaders responded. Nearly every American state maintained institutions for forcibly segregating those suffering from hereditary disabilities and, during the period from 1900 to 1935, thirty-two states enacted compulsory-sterilization laws. Ultimately, more than sixty thousand were sterilized under these laws in the United States—including more than twenty thousand people in California alone.[35] Most were patients or residents at state mental institutions, but some programs reached criminals or epileptics.

Beginning with passage of Germany's Law for the Prevention of Genetically Diseased Progeny in 1933, every Nordic nation adopted eugenic-sterilization legislation. Germany's

law, the most far-reaching anywhere, mandated the sterilization of persons determined by genetic-health courts to suffer
from congenital feeblemindedness, schizophrenia, manic depression, severe physical deformity, hereditary epilepsy,
Huntington's chorea, hereditary blindness or deafness, or severe alcoholism. "We must see to it that these inferior people
do not procreate," the noted German biologist Erwin Baur asserted at the time. "No one approves of the new sterilization
laws more than I do, but I must repeat over and over that they
constitute only a beginning."[36] Some three hundred thousand
persons were sterilized under this law between 1933 and
1939, when it was replaced by a euthanasia program designed
to rid the Fatherland of its mentally handicapped "children."

In 1914, Davenport's Eugenics Record Office proposed a
comprehensive state program designed to sterilize one tenth
of the population every generation. "If the work should be
begun during the present decade, it would, in accordance
with conservative estimates of future population, require the
sterilization of approximately fifteen million (15,000,000)
persons during this interval," a Eugenics Record Office report
explained. "At the end of the time we would have cut off the
inheritance of the present 'submerged tenth,' and would
begin the second period of still more eugenically effective
decimal elimination." Society surely would support the elimination of this "most worthless one-tenth," the publication asserted, and as "public opinion rallies to the support of the
measures, a larger percentage could, with equal safety, be cut
off each year."[37]

Although no such mass program took place in America,
the U.S. Supreme Court upheld the constitutionality of a
model eugenics statute, drafted by the Eugenics Record Office and enacted in Virginia, to sterilize patients and residents
at state mental institutions. "It is better for all the world, if instead of waiting to execute degenerate offspring for crime, or

to let them starve from their imbecility, society can prevent those who are manifestly unfit from continuing their kind," Justice Holmes wrote for the court in 1927. Referring to plaintiff Carrie Buck, her mother Emma, and her infant daughter Vivian, Holmes concluded, "Three generations of imbeciles are enough."[38]

———

These compulsory programs and proposals represented merely the most notorious product of the eugenics movement. Eugenicists also attempted to educate the public about their theories—primarily so that they would adopt eugenic practices voluntarily. The best-selling high school biology textbook of the era in the United States, *A Civic Biology*, featured a section on eugenics that identified mental deficiency, alcoholism, sexual immorality, and criminality as hereditary; offered "the remedy" of sexual segregation and sterilization for these disabilities; and urged students to select eugenically "healthy mates."[39] Some theologically liberal Protestant clerics in Britain and the United States launched an effort to require a certification of eugenic fitness as a prerequisite for a couple's church wedding.

The 1917 motion picture *The Black Stork* offers a telling example of eugenics advocacy from the United States. This full-length feature movie, produced by publisher William Randolph Hearst's movie company and based on the actual practices of an eugenics-minded Chicago obstetrician, explicitly encouraged couples to undergo physical examinations for fitness before marriage and parents to allow their dysgenic newborns to die. It opens with scenes of people who should not breed: a staggering drunk, a street beggar, a boy on crutches, a flirtatious girl in a mental-health institution—all described in the film as "victims of inherited mental or physical disease." Such scenes alternate with ones showing happy youth at play, displaying their healthy physiques. The film's plot centers on a seemingly healthy man who carries a so-called hereditary "taint." He marries without informing his bride about the taint and they have a child who displays the defect at birth. The mother's physician urges her to allow the newborn to die, either by withholding treatment or by lethal medication. She agrees, but only after having a vision of the child's future. In his youth, he is taunted for his limp and hunched back. Sinking into crime and despair as an adult, he ultimately sires a brood of similarly disabled children. "God has shown me a vision of what my child's life would be," the mother exclaims at the movie's climax. "Save him from such a fate," she tells the doctor, who answers this mother's plea by euthanizing her baby.[40]

Eugenics and Social Darwinism lost favor almost as quickly as they had gained it. In reality, they never attracted a widespread public following. Even in its heyday, the popular British essayist G. K. Chesterton dismissed eugenics as a mean-spirited elitist "joke" that briefly "turned from a fad into a fashion."[41] If anything, Social Darwinism attracted even more intense ridicule from its critics. Yet neither encountered much organized opposition, and so a highly motivated eugenics clique could carry the day—for a while. Only the Roman Catholic Church offered sustained, organized opposition to eugenics—it did so largely on religious rather than scientific grounds—and no eugenic-sterilization law ever passed in any jurisdiction where Catholics wielded significant political influence. Labor unions, civil-rights and civil-liberties organizations, and some theologically conservative Protestant denominations occasionally resisted eugenics legislation and Social Darwinian practices that directly affected their interests. Except in Germany, eugenics lacked sufficient depth of support to overcome committed opponents. In the British parliament, for example, one highly critical backbencher, Leonard Darwin's libertarian cousin Josiah Wedgwood, almost singlehandedly stalled passage of eugenics legislation sponsored by his own Liberal Party government in 1912 and 1913.

The tide of popular and expert opinion began turning against eugenics and Social Darwinism by the 1920s. Intelligence tests administered to U.S. Army personnel during World War One showed alarmingly low scores, especially among immigrants from southern Europe. Initially a clarion call for eugenic exclusion, contributing to the enactment of ethnic restrictions on immigration by the U.S. government in 1924, in time these results helped to break the perceived link between low I.Q. scores and degeneracy. Many of the soldiers scoring low on these tests served their country well in war and, presumably, in peace. Further, a new breed of social scientists led by American anthropologists Franz Boaz and Margaret Mead challenged the hereditarian assumptions underlying both eugenics and Social Darwinism. The worldwide economic depression that started during the late 1920s and persisted through the 1930s caused many people to question the simple equation of hereditary fitness with social and economic success. Then the exposé of Nazi practices during the late 1930s and early 1940s turned many prominent eugenicists from championing compulsory sterilization to advocating voluntary birth control. An embarrassed Carnegie Institution finally closed the Eugenics Record Office in 1940. Somewhere in the process, most geneticists quietly climbed off the eugenics bandwagon, as well.[42] The genetic basis of human behavior was simply too complex to address by selective breeding, they now generally conceded. By mid-century, nurture had supplanted nature as the accepted explanation for how people act. Eugenics and Social Darwinism became terms of derision.

CHAPTER 9

AMERICA'S
ANTI-EVOLUTION
CRUSADE

E vangelist Billy Sunday jumped, kicked, and slid across the stage. "Many a minister today has lost his vision. He is standing up in the pulpit preaching tommyrot to the people . . . that we came from protoplasm, instead of being born of God Almighty, instead of being created of the Lord," he shouted in his trademark staccato cadence to a packed house on the first night of his February 1925 Memphis revival. "I don't believe the old bastard theory of evolution. . . . I believe I am just as God Almighty made me." Sweat sprayed from his tossing head as he pounded both fists on the lectern. Sunday probably had given the same sermon at least a hundred times to a total of more than a million people in cities and towns across the United States. He had rehearsed every word and choreographed each gesture.[1]

At the time, Sunday stood alone as the nation's preeminent Protestant evangelist. But his eighteen-day-long Memphis revival held special significance: the Tennessee state senate was then considering legislation banning public-school instruction in human evolution. A senate committee had rejected such a bill prior to Sunday's arrival, but after his various sermons in Memphis drew crowds totaling some two hundred thousand people, the committee reversed itself—leading to enactment of the nation's first law against teaching evolution and a storied showdown over the statute's meaning and validity at the Scopes trial later that year.

Sunday's denunciation of the theory of evolution reflected broad developments within American popular culture. He was not a doctrinaire fundamentalist; Sunday characterized his views as "pure Americanism," which in many ways they were.[2] A product of the nation's rural heartland, he played major-league baseball before feeling the call during the 1890s to become an itinerant evangelist in the tradition of George Whitefield, Charles Gradison Finney, and Dwight L. Moody. Without formal training in theology, Sunday preached the familiar gospel message of individual sinfulness, redemption through personal faith in Jesus, and utter fidelity to the Bible as God's word. He added an earthy, melodramatic style that took the nation by storm in an era of vaudeville theatrics.

Dubbed a "gymnast" for Jesus in his authorized biography, Sunday used slang and stage acrobatics to attract huge crowds in virtually every major American city.[3] Invitations to preach came from a broad spectrum of Protestant churches. Politicians embraced his crusades. Theodore Roosevelt once joined Sunday onstage and Woodrow Wilson invited him to the White House. By his death in 1935, Sunday had preached to more than one hundred million Americans and claimed that more than a million of them had responded to his altar calls.

Sunday opposed evolutionary theories of both human origins and religious understanding. The two blurred in his mind. Embracing developments in biblical higher criticism, many theologically liberal Christians accepted the so-called "modernist" interpretation of the Bible as a collection of accounts about God written over time by various authors, with earlier accounts typically offering more primitive concepts than later ones. Both religious modernism and the scientific theory of evolution denied the literal truth of Genesis, Sunday argued. "When the word of God says one thing and scholarship says another, scholarship can go to hell," he asserted. "If by evolution you mean advance, I go with you, but if you mean by evolution that I came from a monkey, good night!"[4]

When Sunday peopled hell in his sermons, Charles Darwin inevitably flailed in the fiery flames. Huxley and Spencer occasionally joined him. During World War One, German evolutionists, Social Darwinists, and expositors of biblical higher criticism bore the brunt of Sunday's venom. By the time of his 1925 Memphis revival, in the heat of battle over anti-evolution legislation, Sunday focused on evolutionary educators. "Teaching evolution. Teaching about pre-historic man. No such thing as pre-historic man. In the beginning God made man—and that's as far back as it runs," he declared. "A-a-ah! Pre-historic man. Pre-historic man. Ga-ga-ga-ga," at which point, Memphis's leading newspaper reported, "Mr. Sunday gagged as if about to vomit."[5]

———

Although Sunday expressed his opinions more loudly than most conservative Christians, his stated reasons for rejecting Darwinism resonated widely among them. Sunday maintained that any theory of human evolution conflicted with a literal reading of Genesis. Yet no scientist had ever observed people evolving from other primates or, for that matter, one distinctly different kind of animal developing from another. Even if evolution represented the best naturalistic explanation for the origin of species, anti-evolutionists like Sunday declared their intent to stick with a literal reading of God's word until science proved evolution by direct observation. Further, Sunday complained that Darwinism replaced the traditional Christian belief in a perfect original creation broken by human sinfulness with the image of humanity ascending through purely natural processes from savage origins to ever-higher levels of development. The fact that many liberal Christians, spiritual modernists, and agnostics welcomed this reversal of viewpoint made traditional Christians all the more wary. Finally, Sunday linked evolutionary biology to Social Darwinism, eugenics, and other forms of biological de-

terminism that stood in opposition to his message of individual salvation and sanctification available through divine grace
to all people regardless of their supposed genetic fitness.

By the early twentieth century, theologically conservative
Protestants in the United States had splintered into various
subgroups. Evangelicals proclaimed the traditional Protestant
gospel of personal salvation though faith in Jesus and upheld
the Bible as God's inspired word. In the 1910s, a subgroup of
militant evangelicals began calling themselves "fundamentalists" to emphasize their commitment to what they saw as the
fundamental tenets of biblical Christianity: the inerrancy of
Scripture, the veracity of Old and New Testament miracles,
and the trustworthiness of end-time prophecies. Pentecostals
emerged as a separate subgroup claiming power through the
Holy Spirit to heal, prophesy, and speak in tongues. The vast
majority of Americans who identified with these subgroups
shared to some degree Sunday's concerns about the theory of
evolution. Indeed, most conservative Christians never
warmed to Darwinism.

Before Darwin published *Origin of Species* in 1859, orthodox Christians within the scientific community were among
the staunchest defenders of the doctrine of special creation,
and many of them held out the longest against Darwin's ideas.
As scientific support for creationism waned, some theologians, ministers, and lay Christians took up its defense. In his
1874 book *What Is Darwinism?,* for example, the noted Princeton theologian Charles Hodge presented a tightly reasoned
argument leading to the answer, "It is atheism [and] utterly
inconsistent with the Scriptures." Hodge spoke for many conservative Christians when he stressed that Darwin's "denial of
design in nature is virtually the denial of God."[6] Beginning in
the late nineteenth century, conservative Christian publishers
poured forth a steady steam of anti-evolution books and
tracts. In one of his final sermons, Dwight L. Moody damned

the "false doctrine" of materialistic evolution as one critical sin-inducing "temptation" afflicting modern life; after his death in 1899, his ongoing Bible Institute emerged as a center for anti-evolutionism.[7] By the 1920s, many leading American evangelicals and fundamentalists had taken a public stand against the theory of evolution. Powerful Baptist and Presbyterian pastors launched drives to purge denominational colleges and seminaries of Darwinian influences. Among those responding to a 1927 survey of American Protestant ministers, a significant percentage of Lutherans (89), Baptists (63), Presbyterians (35), and Methodists (24) answered "yes" to the question, "Do you believe that the creation of the world occurred in the manner and time recorded in Genesis?"[8]

Notably, only about one in ten of the Episcopalian and Congregationalist ministers responding to this survey affirmed a belief in the Genesis account of creation. Because of their wealth and social standing, Episcopalians and Congregationalists tended to carry weight in elite culture, higher education, and state politics disproportionate to their numbers. Evolutionism often became part of the religious worldview of liberal theologians and ministers in these and other Protestant denominations. The renowned Congregational pastor Henry Ward Beecher blazed the trail in 1885 by publishing *Evolution and Religion,* in which he extolled evolution as "the method of God in the creation of the world" and in the development of human society, religion, and morality. "Evolution is accepted as *the method* of creation by the whole scientific world," Beecher wrote. "It is the duty of the friends of simple and unadulterated Christianity to hail the rising light and to uncover every element of religious teaching to its wholesome beams."[9]

In 1922, the mounting concerns of American evangelicals and fundamentalists erupted into a nationwide effort to drive Darwinism from public education. More than anyone,

William Jennings Bryan transformed an inward-focused campaign to purify church doctrine into an outward-looking crusade to change government policy.

OUTLAWING A THEORY

Bryan was a legend in his own lifetime. A political liberal with decidedly conservative religious beliefs, he entered Congress in 1891 as a young, silver-tongued Nebraska populist committed to defend rural America from economic exploitation by Eastern bankers and railroad barons. Rejecting the Social Darwinian government policies of his day, Bryan delivered his most famous speech at the 1896 Democratic National Convention, where he demanded an alternative silver-based currency to help debtors cope with the crippling deflation caused by reliance on gold-backed money. "You shall not press down upon the brow of labor this crown of thorns," he shouted in an address heard from Wall Street banking houses to Rocky Mountain silver mines, "you shall not crucify mankind upon a cross of gold."[10] The speech electrified the convention and secured him the Democratic presidential nomination; at age thirty-six, he was the youngest person ever so honored by a major political party. A seasoned orator exploiting the nation's new network of railroads, Bryan carried his campaign to the people. More Americans heard him speak during that campaign than had ever heard anyone in so short a period. Bryan became known as the "Great Commoner" and changed how candidates ran for president. Front-porch campaigns gave way to whistle-stop tours.

A narrow defeat against a favored opponent did not diminish Bryan's standing. He secured two subsequent presidential nominations and served as secretary of state in the Wilson administration, all the while denouncing imperialism abroad and exploitive business practices at home. Although he was trained as a lawyer, Bryan's principal vocation became speaking and writing, with his words coming from both the political left and the religious right. During the balance of his life, he delivered an average of more than two hundred speeches a year and wrote dozens of popular books. In the 1920s, Bryan began speaking out against Darwinism with a shrill tone of urgency.

Two decades earlier, Bryan had criticized the theory for the support it gave to Social Darwinism. "The Darwinian theory represents man as reaching his present perfection by the operation of the law of hate," Bryan complained in 1904, "the merciless law by which the strong crowd out and kill off the weak."[11] He said little else publicly about evolution until 1921, when he began blaming a materialistic, survival-of-the-fittest philosophy for both German militarism during World War One and a loss of religious faith among educated Americans.

His standard argument had two prongs. First, he claimed that the theory of evolution was neither scientific nor credible. "Science to be truly science is classified knowledge," Bryan argued. "Tested by this definition, Darwinism is not science at all; it is guesses strung together." He inevitably bolstered this point by ridiculing various evolutionary explanations for human organs—such as the eye, which supposedly began as a light-sensitive freckle. "The increased heat irritated the skin—so the evolutionists guess, and a nerve came there and out of the nerve came the eye! Can you beat it?" Bryan asked rhetorically. "Is it not easier to believe in a God who can make an eye?" Second, he laid out the dangers of accepting such an unproven hypothesis as true. "To destroy the faith of Christians and lay the foundations for the bloodiest war in history would seem enough to condemn Darwinism," he concluded.[12]

Although Bryan spoke out against it, he did not initially call for laws against teaching evolution. That changed in January 1922, after he heard about such a proposal in Kentucky. "The movement will sweep the country, and we will drive Darwinism from our schools," Bryan wrote to the proposal's sponsor. "We have all the Elijahs on our side. Strength to your arms."[13] With a clear legislative objective in sight, the anti-evolution effort became a political crusade. Bryan spent the two months touring Kentucky in support of the proposal, which lost by a single vote in the state's House of Representatives. Teach students that they descended from apes, Bryan told audiences, and they will grow up to act like monkeys.

———

The crusade spread quickly. Protestant ministers and evangelists who had backed efforts to purify their churches of Darwinian influences enlisted in the new push against teaching evolution in public schools. But Bryan remained the principal driver, giving hundreds of speeches, writing scores of newspaper articles, and publishing three popular books on the topic. The timing and intensity of the protest (coming as it did more than sixty years after Darwin published *Origin of Species*) surprised evolutionists. It certainly puzzled Bryan's wife, who privately cautioned her husband against pushing the matter too far. "Just why the interest grew, just how he was able to put fresh interest into a question which was popular twenty-five years ago, I do not know," she commented in 1925. "The vigor and force of the man seemed to compel attention."[14]

Yet even Bryan could not seed a storm on a cloudless day. Undoubtedly the spread of compulsory public secondary education shaped the particular form that anti-evolutionism took in the 1920s. Prior to that time, most Americans did not attend high school and many communities did not provide public education beyond the eighth grade. The expansion of public secondary education carried evolutionary teaching to an increasing number of students, and did so by force of law at taxpayer expense. Thus Bryan could ask, "What right have the evolutionists—a relatively small percentage of the population—to teach *at public expense* a so-called scientific interpretation of the Bible when orthodox Christians are not permitted to teach an orthodox interpretation of the Bible?"[15] The same legislature that passed the nation's first law against teaching evolution created Tennessee's first comprehensive system of state-supported high schools. Tennessee governor Austin Peay believed that he had to accept the former to secure the latter. Whatever underlay its timing, however, the effect of Bryan's crusade was stunning. An editorialist for the *Chicago Tribune* observed, with a mixture of amazement and concern, "William Jennings Bryan has half of the country debating whether the universe was created in six days."[16]

At the time, most American states had part-time legislatures that only met in general session during the first few months of odd-numbered years. Kentucky was an exception, but when its anti-evolution bill died early in 1922, Bryan and his followers had to wait until 1923 for their next shot at lawmaking. The legislatures in six Southern and border states (including Tennessee) actively debated anti-evolution laws during the spring of 1923, but only two lesser measures passed. The Oklahoma legislature barred the purchase of Darwinian textbooks with state funds; Florida's lawmakers adopted a resolution urging public-school teachers not "to teach as true Darwinism or any other hypothesis that links man in blood relationship to any form of lower life."[17]

Sobered by their failures, anti-evolutionists focused their attention on building grassroots support in Tennessee and a few other promising states in advance of the 1925 legislative sessions. Victories in those states then could lead to later successes elsewhere, they reasoned. Bryan, Sunday, and other prominent national anti-evolution leaders spoke in Tennessee on multiple occasions during 1924. Thanks to their efforts, teaching evolution became a major issue during the 1924 elections, with many legislative candidates vowing to support "Bryan and the Bible."

Representative John W. Butler, a farmer-legislator and Primitive Baptist lay leader from rural east Tennessee, offered an anti-evolution bill of his own composition shortly after the Tennessee House of Representatives convened in January 1925. Butler proposed making it a misdemeanor, punishable by a maximum fine of $500, for a public-school teacher "to teach any theory that denies the story of the Divine Creation of man as taught in the Bible, and to teach instead that man has descended from a lower order of animal." Most of Butler's colleagues apparently already agreed with this proposal, because six days later the House passed it without amendment and virtually without debate.

After the lower house acted so quickly and decisively, partisans on both sides focused their attention on the state Senate. Almost overnight, Butler's bill became the subject of petitions, church sermons, and newspaper articles. Educators, editorialists, and liberal clerics tended to denounce the proposal; evangelicals and fundamentalists embraced it. Acting in the glare of publicity, the Senate judiciary committee repeatedly voted down various anti-evolution measures and the full Senate tabled Butler's bill, but Speaker L. D. Hill, a devout Campbellite Protestant, kept the legislation alive until Billy Sunday returned for his second Memphis revival in as many years.

"A star of glory to the Tennessee legislature, or that part of it involved, for its action against that God forsaken gang of evolutionary cutthroats," Sunday told his audience on the first night of the revival—and soon the Senate earned its star, too.[18] During its spirited three-hour floor debate over Butler's bill, few senators addressed the scientific merits of Darwinism. Instead, lawmakers on both sides dwelt on issues of religious freedom. Proponents, including Hill, argued that public schools should not force students to learn theories that undermine their religious beliefs. Opponents countered that no one's religion should set the standards for science education in public schools. One reluctant supporter justified his vote by saying that "an overwhelming majority of the people of the state disbelieve in the evolution theory and do not want it taught to their children."[19] A colleague estimated that majority at 95 percent. Ultimately, the Senate bowed to popular opinion.

Bryan rejoiced upon hearing that Tennessee had outlawed teaching the theory of human evolution. "Other states North and South will follow the example of Tennessee," he predicted.[20] Fearing that result, opponents of the law set about to derail it. Leading this charge, the American Civil Liberties Union (ACLU) issued a press release in New York City offering to defend any Tennessee schoolteacher willing to challenge the validity of the new statute in state court. Its leaders saw the law as a clear violation of free speech, academic freedom, and the separation of church and state: three principles standing at the core of the ACLU's civil-liberties agenda but which, at the time, received scant legal protection against acts committed by state governments. John Scopes, a twenty-four-year-old science teacher in the small east Tennessee town of Dayton, promptly accepted the ACLU's offer.

MONKEY TRIAL

Like so many archetypal American events, the trial itself began as a publicity stunt. Inspired by the ACLU offer, Dayton civic leaders saw a chance to gain attention for their ambitious young community.[21] "The town boomers leaped to the assault as one man," H. L. Mencken reported. "Here was an unexampled, almost a miraculous chance to get Dayton upon the front pages, to make it talked about, to put it upon the map."[22] Scopes became their willing defendant at the urging of local school officials, even though, strictly speaking, he was not a biology teacher. The young teacher was neither jailed nor ostracized. Quite to the contrary; in the month before his trial, Scopes was feted at a formal dinner in New York City; embraced by the presidents of Harvard, Columbia, and Stanford universities; received at the Supreme Court in Washington; and awarded a scholarship for graduate study at the University of Chicago. When it became clear that the ACLU was seeking to discredit Tennessee's new anti-evolution statute through the Scopes trial, Bryan offered to assist the prosecution. If the town boomers of Dayton wanted a show trial, then Bryan would give them one.

To the extent that Bryan then stood as America's foremost champion of Christian government, Clarence Darrow stood as his opposite. Darrow first gained fame during the 1890s as a criminal-defense lawyer for labor organizers and militant leftists. His notoriety grew as he spoke out against religious influences in public life, particularly biblically inspired legal restrictions on personal freedom. His opposition to religious lawmaking stemmed from his belief that revealed religion, especially Christianity, divided people into warring sects and represented an irrational basis for action in a modern scientific age. In speeches and popular books, Darrow sought to expose biblical literalism as foolish and harmful. He offered rational science—particularly an ill-defined Lamarckian form of evolutionism—as a more humane foundation for ethics. When Bryan volunteered to prosecute Scopes, Darrow signed up to defend him. The sixty-seven-year-old trial lawyer immediately became the brightest light in an already luminous defense team assembled by the ACLU to challenge Tennessee's anti-evolution law.

People everywhere called it "the Monkey Trial." News of

it dominated the nation's headlines during the weeks prior, and pushed nearly everything else off American front pages throughout the eight-day event. Two hundred reporters covered the story in Dayton, including some from Europe. Thousands of miles of telegraph wires were hung to transmit every word spoken in court, and pioneering live radio broadcasts carried the oratory to the listening public. Newsreel cameras recorded the encounter, with the film flown directly to major American cities for projection in movie houses. Telegraphs transmitted more words to Britain about the Scopes trial than had ever before been sent over transatlantic cables about any single American event. Trained chimps performed on the courthouse lawn as a carnival-like atmosphere descended on Dayton. The courtroom arguments addressed the nation rather than the jurors. Both sides agreed on one fact: The American people would decide this case.

The defense divided its presentation among its three principal attorneys. The prominent New York attorney Arthur Garfield Hays raised the standard ACLU arguments that Tennessee's anti-evolution statute violated the individual rights of teachers. Bryan's former Assistant Secretary of State, Dudley Field Malone, a liberal Catholic divorce lawyer, argued that the scientific theory of evolution did not conflict with a modernist interpretation of Genesis. Darrow, for his part, concentrated on debunking fundamentalist reliance on revealed scripture as a source of knowledge about nature suitable for setting education standards. Their common goal, as Hays stated at the time, was to make it "possible that laws of this kind will hereafter meet the opposition of an aroused public opinion."[23]

The prosecution countered with a half dozen local attorneys led by the state's able prosecutor and future U.S. Senator Tom Stewart, plus Bryan and his son, William Jennings, Jr., a Los Angeles lawyer. In court, they focused on proving that

Scopes violated the law and objected to any attempt to liti-
gate the merits of that statute. The public, acting through
elected legislators, should control the content of public education, they maintained. The elder Bryan, who had not practiced law for three decades, remained uncharacteristically
quiet in court, and saved his oratory for lecturing the assembled press and public outside the courtroom about the vices
of teaching evolution and the virtues of majority rule.

After the defense lost a pretrial motion to strike the statute
as unconstitutional, the prosecution presented uncontested
testimony by students and school officials that Scopes had
taught evolution. Following this presentation, the defense attempted to offer the testimony of a dozen nationally recognized evolutionary scientists and liberal theologians, all
prepared to defend the theory of evolution as valid science
that could be taught to no public harm. The prosecution immediately objected to such testimony as irrelevant to the
issue of whether Scopes broke the law. The anti-evolution
statute was not on trial, prosecutors argued, only the defendant. After three more days of debate, the judge sided with
the prosecution. The trial appeared to have ended without
ever directly addressing the supposed conflict between evolutionary science and biblical Christianity.

Frustrated by his failure to discredit the law through the
testimony of scientists and theologians, Darrow invited Bryan
to take the stand in its defense. Bryan accepted Darrow's challenge. Up to this point, lead prosecutor Tom Stewart had
masterfully limited the proceedings and confined his wily opponents. But Stewart could not control his impetuous co-
counsel. "They did not come here to try this case," Bryan
explained early in his testimony. "They came here to try revealed religion. I am here to defend it, and they can ask me
any questions they please."[24] Darrow did just that.

Thinking the trial all but over, and hearing that cracks had appeared in the ceiling below the overcrowded second-floor courtroom, the judge had moved the day's session outside, onto the courthouse lawn. The crowd swelled as word of the encounter spread. From the five hundred persons initially in the courtroom, the number rose to an estimated three thousand spread over the lawn—nearly twice the town's normal population. Darrow posed the well-worn questions of the village skeptic: Did Jonah live inside a whale for three days? How could Joshua lengthen the day by making the sun (rather than the earth) stand still? Where did Cain get his wife? In a narrow sense, as Stewart persistently complained, Darrow's questions had nothing to do with the case because they never inquired about human evolution. In a broad sense, as Hays repeatedly countered, they had everything to do with it because they challenged biblical literalism. Best of all for Darrow, no good answers existed. Bryan could either affirm his belief in seemingly irrational biblical accounts, and thus expose that his opposition to teaching about evolution rested on narrow religious grounds, or concede that the Bible required interpretation. He tried both tacks at various times without appreciable success. To Bryan's growing frustration, Darrow never asked about the theory of evolution itself. He knew the Great Commoner would deliver a stump speech in response.

Darrow raised only two issues involving the supposed conflict between science and Scripture, and in both cases Bryan sought to reconcile them. In a modest concession to Copernican astronomy, Bryan suggested that God extended the day for Joshua by stopping the earth rather than the sun—an occurrence that would defy the laws of Newtonian physics, Darrow noted. Similarly, in line with established evangelical scholarship dating back to the days of Georges Cuvier, Bryan affirmed his understanding that the Genesis days of creation represented geologic ages or periods, leading to the following exchange, with Darrow asking the questions:

Courtroom photographs of Clarence Darrow (left) and William Jennings Bryan (right), with coat and collar removed because of the heat, at the trial of John Scopes in Dayton, Tennessee, July 1925.

"Have you any idea of the length of these periods?"

"No; I don't."

"Do you think the sun was made on the fourth day?"

"Yes."

"And they had evening and morning without the sun?"

"I am simply saying it is a period."

"They had evening and morning for four periods without the sun, do you think?"

"I believe in creation as there told, and if I am not able to explain it I will accept it."[25]

The earth could be six hundred million years old, Bryan admitted. Though he had not ventured far beyond the bounds of biblical literalism, the defense made the most of it. "Bryan had conceded that he interpreted the Bible," Hays gloated. "He must have agreed that others have the same right."[26] Of course the reporters loved it. Forget Scopes and his inevitable conviction by a jury that had heard but two hours of testimony during the week-long trial (and none of Bryan's testimony); the lead story became the Great Commoner's public humiliation at the hands of the man Bryan denounced in the midst of his ordeal as "the greatest atheist or agnostic in the United States."[27] A next-day editorial in the usually staid *New York Times* commented about Bryan, "It has long been known to many that he was only a voice calling from a poorly furnished brain-room. But how almost absolutely unfurnished it was the public didn't know till he was forced to make an inventory."[28]

Most neutral observers viewed the trial as a draw, and few saw it as decisive. America's adversarial legal system tends to drive parties apart rather than reconcile them, and that certainly resulted in this case. Despite Bryan's stumbling on the witness stand, both sides effectively communicated their message from Dayton—maybe not well enough to win converts, but at least sufficiently well to energize those already predisposed toward their viewpoints. Due largely to the media's portrayal of Darrow's effective cross-examination of Bryan, later made even more cutting in the popular 1955 play and 1960 movie *Inherit the Wind,* millions of Americans thereafter ridiculed religious opposition to the theory of evolution. Yet the widespread coverage given Bryan's impassioned objections made anti-evolutionism all but an article of faith among conservative American Christians. When Bryan died a week later in Dayton, they acquired a martyr to this cause.

Anti-evolution activism increased following the trial, but it encountered growing resistance. Mississippi and Arkansas promptly passed statutes modeled on the Tennessee law and several other states imposed lesser restrictions. An anticipated legislative victory in Minnesota turned into a demoralizing defeat, however. When one Rhode Island legislator introduced such a proposal in 1927, his bemused colleagues referred it to the Committee on Fish and Game, where it died without a hearing or a vote. A forty-year-long standoff resulted: A hodgepodge of state and local restrictions on teaching evolution coupled with the heightened sensitivity of some parents elsewhere led most high-school biology textbooks and many individual teachers virtually to ignore the subject of organic origins. Consequently, after the Tennessee Supreme Court reversed Scopes's conviction on a technicality in 1927, and when no state or locality brought any other prosecutions under their anti-evolution laws, courts did not have another opportunity to review the meaning and validity of those restrictions until the 1960s. By then, the scientific and religious landscape in America had changed in two key respects. On the one hand, opinion among biologists on how evolution operated coalesced around the starkly Darwinian modern synthesis. On the other hand, opinion among conservative Christians hardened in its fidelity to the biblical account of creation. These developments took decades to unfold, however. For the time being, America's anti-evolution crusade had run its course.

CHAPTER 10

THE MODERN SYNTHESIS

It was called the "peppered moth" because of its appearance: mostly white with a distinct sprinkling of black specks on the back and wings. Supposedly its speckled appearance, which blends with the lichen-covered tree branches of rural England, helped camouflage it from predators during its daylight resting hours. No one knows when and where the first black specimens appeared, but tradition places that storied event at or around 1848 in the vicinity of Manchester, England. At the time, Manchester was evolving from an agricultural market town into a major industrial city. Soot from countless coal-fired mills and stoves blackened tree bark, walls, and other surfaces for miles around the city. Industrial pollution killed the lichens. Gradually the speckled moths came to stand out against the grimy landscape while the black ones blended into it. The peppered moth evolved along with its environment. In 1896, naturalist J. W. Tutt reported that 98 percent of peppered moths in and around Manchester were now black. He found similarly high percentages of the dark variety in other English industrial regions, but virtually none in rural counties. Tutt and other late-Victorian naturalists hailed it as an example of evolution in action.

Attributing the transformation to evolution did not resolve the debate over how the process worked. Lamarckians attributed the change to acquired color characteristics. Theistic evolutionists saw God at work. Mutation theorists chalked it up to the widespread occurrence of similar mutations under stressful conditions or to the spread of a favorable mutation with scant contribution from natural selection. Biometricians saw it as a product of the natural selection of continuous variations within the population.[1] By the turn of the twentieth century, no consensus existed among biologists about how evolution operated. Even the rise of Mendelian genetics did not settle the matter, because early geneticists tended to see the new science as a refinement of mutation theory (with genes simply the material locus for hereditary mutations transmitted in Mendelian ratios), without revising their understanding of evolution to incorporate a fundamental role for natural selection. Especially in Britain, biometricians distanced themselves from Mendelism as much as from mutation theory. In the 1920s, the mathematically gifted British biochemist J.B.S. Haldane drew on the familiar example of peppered moths to begin, concurrently with Ronald A. Fisher in England and Sewall Wright in America, linking biometry to Mendelism through the study of how genetic changes affect populations, or "population genetics."

Perhaps it helped that Haldane was instinctively contrary. The only son of a union between two aristocratic Scottish families, with a father renowned as a physiologist and a passionately feminist mother, Haldane was precocious, strong-willed, supremely self-confident, and given to bullying both subordinates and superiors. Where his father once delivered the famed Gifford Lectures, founded to promote natural theology, and publicly declared, "This is a spiritual world," the son proclaimed his faith in dialectic materialism and joined the Communist Party. Where his uncle once served as Britain's war minister, Haldane ultimately resigned his chair in biology at University College, London, ostensibly in protest of post–World War Two British militarism, for an academic position in the newly independent, purportedly pacifistic nation of India. Long before he left Britain, however, Haldane made his mark in genetics, biometry, and evolution theory. His prolific work consistently presented a strictly naturalistic view of the origins and development of life on earth.

In a series of ten highly mathematical papers published between 1924 and 1934, Haldane sought to show that the natural selection of genetic variations transmitted in Mendelian ratios could produce adaptive change in populations. In short, he argued, Darwinism plus Mendelism equals evolution. This idea was not new. As early as 1902, shortly after the rediscovery of Mendelism, British statistician G. Udny Yule had suggested something of the sort, but bitter personal, professional, and philosophical rivalries between Mendelians and biometricians delayed the idea's development for two decades. Reconciliation came through mathematical evidence showing that varieties enjoying even a slight competitive advantage would, over multiple generations and in a process acting like compound interest in banking, come to predominate within a population. "A satisfactory theory of natural selection must be quantitative," Haldane wrote in the opening lines of his initial 1924 paper. "In order to establish the view that natural selection is capable of accounting for the known facts of evolution we must show not only that it can cause a species to change, but that it can cause it to change at a rate which will account for present and past transmutations."[2]

Haldane was a theoretical biologist. Due to what his biographer referred to as his "ham-handedness," coupled with his impatient temperament, he never conducted meticulous field or laboratory research in genetics.[3] Instead, in his 1924 paper, he used Tutt's field data on peppered moths to illustrate how selection for a genetic variation (such as the moth's black color) could feed rapid evolutionary change in an environment favoring that variant. The observed increase in the percentage of black specimens within the peppered-moth population around Manchester—from 1 percent in 1848 to 99 percent in 1898—required only a 50 percent higher sur-

vival rate for black moths over speckled ones, Haldane calculated, which was clearly plausible, given the changed environment. In contrast, he noted, explaining the increase solely by
individual variations without selection (as early Mendelians
tended to do) would require one in five moths to mutate from
speckled to black—an obvious impossibility. "The only probable explanation," Haldane concluded, "is the not very intense degree of natural selection postulated."[4]

In this and other papers, Haldane made similar computa-
tions testifying to the adaptive power of Darwinian selection
at work in Mendelian populations. He did not prove that natural selection drives the evolutionary process; indeed, the
peppered-moth example did not even involve the evolution
of a new species. He did, however, help revive scientific interest in selectionism under the assumption that whatever
caused minor variations within a local population, by extrapolation, also caused major ones at the species level and above.
By 1932, Haldane could open his book *The Causes of Evolution*
by mocking the allegedly popular refrain "Darwinism is
dead." The eclipse of Darwinism that had spread across evolutionary thought during the preceding generation had
passed, he asserted, "mainly due to R. A. Fisher, S. Wright,
and myself."[5]

———

Unlike Haldane, Fisher had only one species in mind
when he made his groundbreaking statistical contributions to
the modern understanding of natural selection: humans.
While Haldane believed in eugenics, and suggested in *The
Causes of Evolution* that only "the best thousandth of the present human race" should reproduce, the subject utterly consumed Fisher, who became one of its last great scientific
champions.[6]

Born into a prosperous English family that fell on hard
times during his youth, Fisher attended Cambridge Univer-

Ronald A. Fisher, with calculator, in his Cambridge University office, 1952.

sity on scholarship. There he displayed two traits that would shape his professional and personal life—a stunning facility for mathematics and a brooding preoccupation with breeding better Britons. Fisher loved England and wanted its people to remain vigorous and strong. His patriotism turned him to eugenics, and thence to population genetics. As a college student, Fisher recognized that the theoretical reconciliation of Mendelism with biometry lay in mathematically distinguishing hereditary elements of variation (all of which he attributed to genetic factors) from nonhereditary ones (which he ascribed to environmental factors), and statistically understanding how multiple genes affect the former to produce the seemingly continuous array of variations found in natural populations.

Socially inept, careless about his personal appearance, slight in stature with an oversized head, and rejected for military service during World War One due to his extremely poor eyesight, Fisher set about to save the English people from themselves by computing the mathematics of selectionism to show that eugenics works. On a personal level, he overcame his extreme discomfort with women enough to do his eugenic duty for England by marrying and siring nine children, though years of neglect and abuse eventually drove his wife to divorce him. Fisher outlined his eugenic vision in a 1911 student paper delivered to his collegiate eugenics club. He noted that if twenty gene pairs contributed to intelligence, then on average the best combination of them produced by random mating would occur in the English population only once in more than twenty thousand generations. "It will give some idea of the excellence of the best of these types when we consider that the Englishmen from Shakespeare to Darwin (or choose who you will) have occurred within ten generations," Fisher observed. "The thought of a race of men combining the illustrious qualities of these giants, and breeding true to them, is almost too overwhelming, but such a race will inevitably arise in whatever country first sees the inheritance of mental characters elucidated."[7]

Like his hero Francis Galton, whose endowed professorship in eugenics at University College, London, he assumed in 1933, Fisher's concern for eugenics spurred him to make fundamental contributions to theoretical mathematics and the quantification of evolutionary biology. He was the finest statistician of his day, and the practical motivations of his work have not diminished its scientific significance. Beginning with a 1918 paper on correlations among relatives and culminating in his 1930 book, *Genetical Theory of Natural Selec-*

tion, Fisher showed that a Darwinian selection process acting on a large, genetically varied population subject to Mendelian laws of inheritance favored the diffusion of beneficial genes. The greater the benefit conferred by these genes in a given environment, the faster their frequency would increase within the population, he calculated. Change the environment such that different genes conferred benefits, and gene frequencies would shift accordingly. Extrapolated to the species level, this view holds that evolution acts through gene selection in a virtually continuous process that finely adapts organisms to their environment. In his publications, Fisher stressed the eugenic implications of his theoretical insights. Perhaps reflecting his own straitened financial situation, he particularly touted state-funded family allowances that rose with income and family size so as to encourage upper-middle- and professional-class couples to have many children. Although the British government never implemented his elitist public-policy proposals, it knighted him in 1952 for his service to British science.

Thanks to the work of Fisher, Haldane, and others, by 1930 biologists began appreciating the genetic complexity of large populations. Early geneticists had studied discontinuous, single-gene traits (such as tall versus short pea plants), thinking that single-gene mutations gave birth to new species or varieties. They downplayed the seemingly continuous variability within large natural populations (such as the variable heights of people). As the discipline matured, geneticists recognized that multiple-gene interactions affect traits such that individual gene changes could cause seemingly continuous variations. For example, the Swedish biologist Herman Nilsson-Ehle calculated that if ten different genetic factors affect a trait, then sixty thousand variations of it might exist. Such thinking eventually displaced the blending view of inheritance that had dominated biology for generations and persisted in the biometric interpretation of continuous variation. Under Mendelism, favorable variations could spread through a population without risk of swamping and unfavorable ones could survive if recessive. Through genetic recombination and mutation, large populations could build a fund of variability available for later selection should environmental conditions warrant.

———

The emerging viewpoint recognized the adaptive power of natural selection to drive the evolution of large, genetically varied populations in a continuous direction, such as toward darker peppered moths or bigger-brained hominids. Classic Darwinism spoke of a branching pattern of evolutionary development, however, as suggested by the different finch species scattered across the Galápagos Islands. Unlike laboratory geneticists and theoretical biologists, field naturalists typically confronted this branching pattern in their research—such as when they encountered small pockets of closely related species existing in isolation on the fringes of a main population. Lacking a Darwinian explanation for this form of evolution, field naturalists typically clung to Lamarckism longer than biologists from other disciplines. It took the work of Sewall Wright to begin bringing them into the Darwinian fold.

Although historians of biology rightly lump Wright with Fisher and Haldane as the co-founders of population genetics, Wright brought fundamentally different expertise and insight to the field. They were primarily mathematicians who constructed statistical models; he was a physiological geneticist with a background in animal-breeding research and a learned facility with mathematics. They stressed the evolutionary significance of large, genetically varied populations; he focused on the contribution of small, genetically restricted

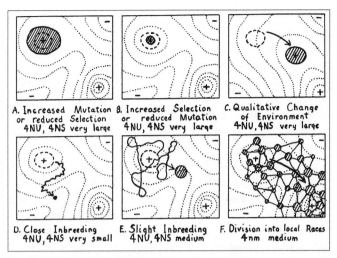

A. Increased Mutation
or reduced Selection
4NU, 4NS very large

B. Increased Selection
or reduced Mutation
4NU, 4NS very large

C. Qualitative Change
of Environment
4NU, 4NS very large

D. Close Inbreeding
4NU, 4NS very small

E. Slight Inbreeding
4NU, 4NS medium

F. Division into local Races
4nm medium

Sewall Wright's 1932 diagram of the adaptive landscape of genetic fitness, allowing the evolution of organisms due to mutations, selection, environment, population size, and crossbreeding.

ones. They thought selection acted on individual genes; he saw it operating on complex genetic interactions as expressed in organisms. Wright also left the field with its most powerful metaphor—that of an adaptive landscape. At least in part because many biologists could not understand the mathematical models offered by Fisher and Haldane (or by Wright, for that matter), this simple metaphor for how species evolve made a lasting impact on an entire discipline.

Consider a landscape with hills and valleys, Wright suggested in a 1932 paper. Each point on the surface represents a possible type of organic population, with the most similar types (differing perhaps only in one gene or genetic interaction) located next to each other and more different ones farther away. The elevation of each point on the surface reflects its Darwinian fitness, with greater fitness indicated by greater height. The surface should flow smoothly up and down, because minor genetic variations should only marginally affect fitness. Natural selection acting on random genetic variations should drive populations up toward peaks of fitness, Wright noted, but could not fully account for one species branching out into many. Branching would require subpopulations of organisms to travel down from their current peaks of fitness, across valleys of relative unfitness, and back up other peaks of fitness—all through a process of incremental genetic variation.

Wright envisioned the process involving a subpopulation of a species becoming isolated from the main population, perhaps on the geographic fringe of the species' range. If the subpopulation were small enough and subject to intense inbreeding (which stimulates genetic interactions and brings out recessive traits), then selection might not operate to maximize its adaptive fitness. In his metaphor, the subpopulation would move downhill and begin wandering across the valley. Wright called the phenomenon "genetic drift." If the subpopulation survived, then random changes in its genetic makeup might carry it toward a new adaptive peak. Natural selection would then drive it up the peak toward greater fitness, resulting in a new species. If the newly occupied peak stood higher than the old peak, then the population's new form might supplant its progenitor in the ensuing struggle for existence. According to Wright's theory, genetic drift functioned in a "shifting balance" with natural selection to generate new species through alternating periods of genetic restriction (or "bottlenecks") and expansion. The entire process worked through "trial and error." In the real world, this theory suggested that small, isolated subpopulations constituted the seeds of new species—just as field naturalists suspected all along.[8]

Although field naturalists found his theory attractive, Wright discovered it through his work with selective breeding. Beginning with his position at the U.S. Department of Agriculture from 1915 to 1925, and throughout his subsequent career at the University of Chicago, Wright studied the effects of inbreeding on guinea pigs. At the Department of Agriculture, he also studied the cultivation of domesticated shorthorn cattle. In both instances, he demonstrated that the establishment of new varieties could be accomplished through isolating a small subpopulation with a particular trait (such as short horns), fixing that trait through inbreeding, and then introducing it into the main population. This, he argued, was how new species often arise in the wild.[9]

Fisher objected to Wright's reliance on nonadaptive mechanisms in evolution. Wright countered that Fisher "overlooks the role of inbreeding as a factor leading to nonadaptive differentiation of local strains, through selection of which, adaptive evolution of the species as a whole may be brought about more effectively than through mass selection of individuals."[10] They agreed that mass selection pushes large populations toward adaptive peaks, but Wright believed the process would stagnate under static conditions. Fisher shot back, "Static conditions in the evolutionary sense certainly do not occur, for...the evolutionary progress of associated organisms ensures that the organic environment shall be continentally changing."[11] It came down to opposing examples of inbred shorthorn cattle versus environmentally adapted peppered moths. The ongoing dispute helped to supply population genetics with the rich diversity of mathematical models and mechanical metaphors needed to finally displace Lamarckian and other vitalist concepts from biology.

———

Among the field naturalists attracted by Wright's adaptive-landscape metaphor, none proved more influential than Theodosius Dobzhansky, a Russian émigré who had moved to America in 1927. Largely invisible to outsiders after the Bolshevik Revolution in 1917, genetics in Russia developed along lines parallel to those in Britain and the United States—and continued to do so until the late 1930s, when iron-fisted state support for the Lamarckian ideologue Trofim Lysenko choked off other alternatives. Indeed, had his work been known outside Russia at the time, Sergei Chetverikov would have become a co-founder of population genetics. Chetverikov pioneered the idea that recessive mutations create hidden reservoirs of genetic diversity within populations on which selection can act when conditions warrant. Although he did not study under Chetverikov, Dobzhansky later recalled inspiring visits to the senior genetist's Moscow lab during the mid-1920s.[12] As a result, Dobzhansky came to the United States with a knowledge of population genetics layered atop a field naturalist's familiarity with the wide variability of wild populations. Working in America first with Thomas Hunt Morgan and later as a professor in California and New York, Dobzhansky developed an understanding of how natural populations evolve that drew heavily on Wright's shifting-balance approach to the origin of species. He first encountered this approach at the 1932 genetics congress at which Wright introduced the adaptive-landscape metaphor, and (in Dobzhansky's words) it was "love" at first sight.[13]

In his landmark 1937 *Genetics and the Origin of Species,* which translated the abstract theorems of population genetics into a working synthesis of how genetic variations feed the evolutionary process, Dobzhansky laid out his thoughts in terms of an adaptive landscape navigated by a shifting balance of small-scale genetic drift and large-scale selectionism. "Wright considers the situation that may present itself in a species whose population is subdivided into numerous isolated colonies of different size," Dobzhansky wrote. "Such a situation is by no means imaginary; on the contrary, it is very frequently encountered in nature. . . . A subdivision of the species into isolated populations, plus time to allow a sufficient number of generations to elapse . . . is all that is necessary for race formation."[14]

In the first edition of his book, Dobzhansky presented the origin of species in terms of race formation writ large, with both steps involving a mix of adaptive and nonadaptive genetic forces. In later editions, he adjusted the fulcrum of the shifting balance in favor of selection by restricting the range of nonadaptive genetic drift to local populations—an adjustment made largely in response to theoretical developments by Fisher, E. B. Ford, and other British geneticists who continued to press the case for competition in the evolutionary process. In his own response to such evidence, Wright also moderated his position. Iconoclastic evolutionist Stephen Jay Gould later saw these adjustments as evidence that the modern evolutionary synthesis "hardened" over time. Initially the synthesis simply proposed that small-scale, gene-level variations as observed in laboratories and field studies (so-called "micro-evolution") can account for large-scale evolutionary developments (or "macro-evolution"), Gould explained. Ultimately it asserted that selection drives the entire process.[15] In time, the synthesis became known as "neo-Darwinian" precisely because it combined unadulterated Darwinian selection mechanisms with the findings of modern geneticists.

Throughout the various editions of *Genetics and the Origin of Species* and in his other publications, Dobzhansky consistently highlighted the impact of both geographic or other isolating factors and of variability hidden in recessive genes. Indeed, as the synthesis became more selectionist, Dobzhansky labored to show that having two different alleles in a gene locus ("the heterozygote state") rather than two identical ones ("the homozygote state") benefited both the individual (through genetic interactions manifested in its lifetime) and the population (by increasing genetic diversity). Elucidating these factors, together with restating the findings of population genetics in nonmathematical terms and tying them to wild populations, became his book's contribution to the modern evolutionary synthesis. "The reason why the book had whatever success it did was that, strange as it may seem, it was the first general book presenting what is nowadays called...'the synthetic theory of evolution,'" Dobzhansky commented long after his fame was secure. "People who contributed most to it I believe were R. A. Fisher, Sewall Wright, and J.B.S. Haldane; their predecessor was Chetverikov. What that book of mine, however, did was, in a sense, to popularize this theory. Wright is very hard to read."[16]

Dobzhansky contributed much to the rise of the modern neo-Darwinian synthesis in his own research and writing. He also actively recruited others to the cause and worked with them to institutionalize their approach to the study of evolution within the mainstream scientific community. In America, three noted scientists became principals in the process: field zoologist Ernst Mayr, vertebrate paleontologist George Gaylord Simpson, and plant geneticist G. Ledyard Stebbins, Jr. Inspired by Wright's adaptive-landscape metaphor and guided by Dobzhansky's book, each labored to bring his discipline within the fold of modern neo-Darwinian thought through research, writing, and reinterpretation of the evidence. In Britain, T. H. Huxley's grandson, the evolutionary biologist and popular-science writer Julian Huxley, added his own complementary survey of developments in the field with his 1942 book, *Evolution: The Modern Synthesis.*

Building on Dobzhansky's work, Mayr's 1942 book, *Systematics and the Origin of Species,* defined species in synthetic terms as "groups of actually or potentially interbreeding natural populations, which are reproductively isolated from other such groups."[17] Forget morphologic similarities or metaphysical forms; a species became simply a breeding population, however varied. Indeed, Mayr stressed that variation characterizes all nature and that life everywhere is emergent. A new species arises when a geographically separate subpopulation develops such distinctive genetic traits that its members no longer can or will crossbreed with members of its parent population, he explained: It is merely a passing phase in life's continuum of change. To the extent that selection molds these new traits, subpopulations and species are tailored to their particular environments through a process Mayr called "adaptive radiation." If geographic barriers then collapse, competition between related species for food or other resources would push them toward even greater differentiation. Mayr concluded, "A new species develops if a population which has become geographically isolated from its parent species acquires during this period of isolation characters which promote or guarantee reproductive isolation when the external barriers break down."[18] Seen in this way, a species becomes a range of gene frequencies within a reproductively isolated population. Evolution occurs whenever the range shifts. It is a purely materialistic chemical process.

Simpson and Stebbins faced the challenge of bringing the distinguishing features of their disciplines within the domain of an evolutionary theory developed with living animals in mind—but it had to happen in order for the synthesis to become truly comprehensive. Paleontologists saw discontinuous but seemingly linear patterns in the fossil record; botanists witnessed widespread breeding across species lines and apparent evolutionary jumps caused by the polyploid multiplication of chromosomes during reproduction. In his 1944 *Mode and Tempo in Evolution*, Simpson reinterpreted the

Ernst Mayr in the field at 50, 1954.

fossil record to de-emphasize its stops, starts, and dead ends. To explain its gaps, he noted that genetic drift and other forms of rapid evolution within small populations could cause major changes without leaving discernible fossil evidence. Although the fossil record could not prove the modern synthesis, Simpson argued that it was fully consistent with the concept. Stebbins went further. In his 1950 *Variation and Evo-*

lution in Plants, he integrated hybridization and polyploidy
into his overall synthetic conclusion: For plants as for animals,
"individual variation, in the form of mutation (in the broadest
sense) and gene recombination...is sufficient to account for
all the differences, both adaptive and nonadaptive, which
exist between related races and species." Stebbins then added
the essential extrapolation: "The differences between genera,
families, orders, and higher groups of organisms...are similar enough to interspecific differences so that we need only to
project the action of these same known processes into long
periods of time to account for all of evolution."[19]

In their foundational publications, Dobzhansky, Mayr,
Simpson, and Stebbins utilized the adaptive-landscape
metaphor to explicate the evolutionary process. Through it,
they taught their students and their colleagues about life. By
the widely celebrated centenary of Darwin's *Origin of Species*
in 1959, the modern synthesis had become virtual dogma
within biology and its leading proponents sat atop the profession in chaired professorships at elite universities and on the
boards of all the relevant scientific societies. Most crucially,
the synthesis generated a seemingly endless stream of
testable deductions about how populations should act under
controlled conditions and in the wild. Time and again the
theory passed these tests, leading to Dobzhansky's famous
aphorism that "nothing in biology makes sense except in the
light of evolution."[20]

———

Compelling evidence for the synthesis soon came from a familiar source: the striking diversity of finches on the Galápagos Islands. The birds that helped to turn Darwin's thoughts
toward evolution in the 1830s but then sank from prominence
in science became, by the 1950s, the best-known example of
neo-Darwinian evolution in the world. Their transformation
told the story of the modern synthesis in a nutshell (or, more

precisely, a seed casing). Prior to the synthesis, the origins of so many different finch species on one oceanic archipelago remained as much a mystery to biologists as the mechanics of evolution generally. Clearly they must have developed from one immigrant type, evolutionists agreed, but no one could devise a consensus explanation for how it happened.

Between Darwin's day and the emergence of the modern synthesis, a steady stream of scientific expeditions had visited the Galápagos Islands to collect plant and animal specimens. By 1930, major natural-history museums in San Francisco, New York, and London contained thousands of Galápagos finch specimens, yet the sheer magnitude of those collections simply reinforced these birds' puzzling diversity. No ornithologist could impose order on the group. The specimens clearly clustered into a dozen or more morphologic types separated mostly by their beak shapes and body sizes, but the boundaries between species blurred and intermediate forms abounded. The main branches of Galápagos finch evolution seemed obvious enough: Seed, fruit, and insect eaters, each with markedly differing beaks, had developed to exploit different niches in an isolated environment largely devoid of other land birds. What befuddled ornithologists was the persistence within each branch of multiple species, each with its distinctive beak, that apparently overlapped in diet and geographic range. Natural selection should have settled on the fittest beak, Darwinists assumed, and eliminated the others.

In the last major study of these birds published before the modern synthesis became widely accepted, American ornithologist Harry S. Swarth wrote in 1931, "Such remarkable extremes of variation in bill structure as are seen ... lie outside my experience with any North American mainland bird."[21] Expressly rejecting both Darwinian and Lamarckian explanations for this diversity, Swarth divided the finches into forty species and subspecies whose origins he tentatively attributed to the group's extreme innate variability displayed in a uniform environment without sufficient competition to eliminate intermediate types. Responding to Swarth's analysis, British ornithologist Percy R. Lowe suggested instead that the "bewildering diversity, intergradation, and distribution" of finches came from nonadaptive crossbreeding that somehow produced healthy hybrids at a rapid rate unique among birds.[22]

Both these respected, museum-based ornithologists recognized that any final resolution of the origins of these finches required field study of their breeding and feeding habits on the Galápagos archipelago, which was then an isolated Ecuadorean colony populated by a few hundred settlers. "There is no group of birds in the whole world which has more right to occupy the attention of zoologists at the present moment," Lowe asserted at the festivities in London marking the hundredth anniversary of Darwin's 1835 visit to the islands. Rechristening them "Darwin's finches," he urged that "properly qualified investigators...be sent to the Galapagos with the sole object of studying on the spot and for a sufficiently long period, by means of *actual breeding experiments*, the genetics of this very interesting group of birds."[23] Julian Huxley, then secretary of the Zoological Society of London, knew just the person for this job—David Lack.

A twenty-five-year-old English schoolteacher and amateur bird watcher, Lack had come to Huxley's attention as a gifted observer of avian behavior. Huxley arranged for the Zoological Society to pay Lack's expenses for a bare-bones expedition to observe finch behavior on the Galápagos over the course of an entire breeding season. Departing in 1938, Lack traveled by commercial steamers and stayed with local settlers. "The Galápagos are interesting, but scarcely a residential paradise," he noted. "The biological peculiarities are offset by an enervating climate, monotonous scenery, dense thorn scrub, cactus spines, loose sharp lava, food deficiencies, water shortage, black rats, fleas, jiggers, ants, mosquitoes, scorpions, Ecuadorean Indians of doubtful honesty, and dejected, disillusioned European settlers."[24] Money was tight and food insufficient. Despite his personal discomforts (or perhaps because of them), Lack did see something on the Galápagos that no one had seen before—natural selection at work among its finches through interspecies competition.

This observation took time. He saw nothing but co-mingled finches during his four-month stay on the Galápagos, but he did establish that they constituted only thirteen distinct species that did not readily crossbreed. Then he spent more months studying preserved specimens, first in San Francisco and then in New York, where he roomed with the American Museum of Natural History collection's curator, Ernst Mayr. In his initial attempt to interpret his data, published in 1940, Lack shoehorned the finches into the emerging modern synthesis by attributing their evolution to the nonadaptive genetic drift of small, isolated populations.[25] This answer did not fully satisfy him, however. As the synthesis hardened, Lack revised his views to fit its increasing reliance on selection. In doing so, he drew on a rule recently set forth by Russian biologist G. F. Gause: "If two or more nearly related species live in the field in a stable association, these species certainly possess different ecological niches."[26] Since three species of seed-eating ground finches lived together on some Galápagos islands, Lack now asserted, "There must be some factor which prevents them from effectively competing."[27] Their beaks held the key. Although he failed to recognize it while visiting the archipelago during the breeding season, a period in which rain falls and food abounds, Lack ultimately decided that the distinguishing beaks of these ground finches must adapt them for eating different seeds (at least during the dry season). He applied similar analysis to other overlapping species of Galápagos finches.

A complete picture of finch evolution emerged in Lack's 1947 book, *Darwin's Finches.* "That Darwin's finches are so highly differentiated suggests that they colonized the Galapagos considerably ahead of the other land birds," Lack wrote. "The absence of other land birds... allowed them to evolve in directions which otherwise would have been closed to them." In a textbook example of adaptive radiation, some island finches remained ground feeders, others took to the trees for food, and yet others adopted the feeding habits of warblers or woodpeckers. Geographic isolation created island variations among these basic types. But these finches did not remain isolated on separate islands. They spread into one another's territory. "When two related bird species meet in the same region, they tend to compete, and both can persist there only if they are isolated ecologically either by habitat or food," he postulated. Such encounters further pushed the evolutionary development of competing species until they diverge into noncompeting ones, such as by fine-tuning their beaks for different foods. "The evolutionary picture presented by Darwin's finches is unusual in some of its details, but fundamentally it is typical of that which I believe to have taken place in other birds," Lack stated. "So that with these birds, as Darwin wrote, we are brought somewhat nearer than usual 'to that great fact—that mystery of mysteries—the first appearance of new life on this earth.'"[28]

Lack's interpretation of Darwin's finches became the classic case of evolution in action by the 1950s, and prime evidence for the modern synthesis. Featured in countless biology textbooks, nature documentaries, and popular science books, Darwin's finches became practically synonymous with Darwinism, even though Darwin never actually mentioned them in *Origin of Species*. Ornithologists followed Lack to the islands to test and extend his conclusions, most notably the British-born researchers Peter and Rosemary Grant. Beginning in 1973 and still continuing, theirs became the most influential field study of evolution in action ever conducted.

The Grants confirmed Lack's suppositions and nested Darwin's finches even more neatly into the modern synthesis. They observed that bigger-beaked finches eat harder seeds such that, when drought disproportionately decreases the supply of softer seeds, average beak size increases through the mass starvation of smaller-beaked finches. The inverse happens in times of plenty. And always, they found, competition drives apart the average beak size of similar species living together as compared with those same species living apart. In Darwin's finches, beak size is affected by a mix of genes built up over generations though mutations, recombinations, and occasional crossbreeding, the Grants noted. Anything that selects for survival within that mix shifts gene frequencies in the population. "Evolution occurs when the effects of selection on a heritable trait in one generation are transmitted to the next generation," they wrote.[29] At some level, evolution continually happens among Darwin's finches. Projecting from the effects of one severe drought that he witnessed, Peter Grant added, "If droughts occur once a decade, on average, repeated directional selection at this rate with no selection in between droughts would transform one species into another within 200 years."[30]

From the shift of gene frequencies within populations through the origin of similar species to the divergence of biologic kingdoms, modern neo-Darwinian theory relies on the cumulative selection of favorable genetic variations over innumerable generations to account for life's diversity. Darwin's finches represent one particularly well-documented example of the process functioning at lower levels. Studies of other species and groups show similar results. Such evidence breathes life into the mathematical abstractions of population genetics, and gives them added meaning. On the Galápagos Islands, Lack and the Grants found the evolutionary tree of life growing precisely where Darwin first encountered it. Through their work, Darwin's finches joined the peppered moth as mascots for the modern synthesis.[31]

CHAPTER 11

MODERN CULTURE WARS

The late 1950s saw a celebration of Darwinism. With scientists all but agreed on how evolution operated, its study gained standing in science. First Ernst Mayr in 1953 and then George Gaylord Simpson in 1959 moved from museum positions in New York to named professorships at Harvard. Wright, Dobzhansky, and Stebbins advanced within academia. Even David Lack, with no graduate training, obtained a faculty position at Oxford. The American government expanded funding for research on evolution and launched the Biological Sciences Curriculum Study (BSCS), which produced the first widely adopted high school biology textbooks since the Scopes era to employ evolution as an organizing principle. Then came the centennials of the Darwin-Wallace papers in 1958 and *Origin of Species* in 1959. Books and articles on Darwinism marked these occasions for the public. Scientific associations commemorated them with conferences hailing the founders of population genetics and the modern synthesis. Among these major symposia, Haldane presided over an international conference in Singapore; Huxley gave a keynote address at the University of Chicago; and Fisher, Haldane, Huxley, Mayr, and Simpson received special medals at festivities in London. For Fisher and Haldane, these events served as something of a final curtain call—for the others, they provided impetus for continued achievement.

As leading neo-Darwinian scientists with the inclination and ability to write for nonscientists, Huxley and Simpson became two of the best-known scientific intellectuals of their day. Their books sold millions of copies, and the centenary of

Darwinism marked the height of their popularity. Writing for general audiences, Huxley and Simpson reached beyond the formal boundaries of science to address the implications of modern neo-Darwinism for individual ethics, human society, and traditional religion. Thoroughly naturalistic, triumphalist in tone, invoking the authority of science, their messages helped ensure that public controversies surrounding Darwinism would continue under the modern synthesis.

Two decades earlier, Huxley had given up his own scientific research for a public life promoting his vision of "evolutionary humanism." He held a succession of highly visible administrative posts, culminating in his role as the founding Director-General of UNESCO, following World War Two. Huxley used these posts as pulpits to preach his Darwinian gospel of progress.[1] Branching out into new forms of mass communication, he joined the BBC's "Brain Trust," a popular television program whose panelists fielded obtuse questions from the general public in prime time; he also supervised the production of nature films for an English movie company, winning an Oscar for his efforts. Having a celebrated grandfather and two famously brilliant brothers (novelist Aldous Huxley and Nobel laureate for medicine Andrew Fielding Huxley) simply added to his luster. Indeed, Huxley gained such a reputation for intellect that one public-opinion poll ranked his as one of the five "best brains" of Britain. His compulsive publishing, lecturing, and media appearances served a larger mission. Huxley believed in human progress with a religious zeal, and he wanted others to share his belief. For him, progress was embodied in the evolutionary process that gave us birth and would carry us ever upward if we let it.[2]

Unlike his grandfather, who eschewed religion and saw evolution as directionless change that people should resist in their pursuit of human values, Julian Huxley championed a nontheistic "religion without revelation" that worshiped natural selection as its guiding process.³ Although evolution once operated blindly, he believed that humans (having acquired consciousness) can and should channel the selection process to enhance the quality of life for all. "Man's destiny is to be the sole agent for the future evolution of this planet," Huxley told scientists at the 1959 Chicago conclave, "and he must face it unaided by outside help. In the evolutionary pattern of thought there is no longer either need or room for the supernatural. The earth was not created: It evolved. So did all the animals and plants that inhabit it, including our human selves, mind and soul as well as brain and body. So did religion." For Huxley, the evolutionary ethic meant protecting the environment, embracing cultural diversity, and valuing quality over quantity in personal consumption and human reproduction. "Instead of worshiping supernatural rulers, it will sanctify the higher manifestations of human nature," he concluded, "and will emphasize the fuller realization of life's possibilities as a sacred trust. Thus the evolutionary vision, first opened for us by Charles Darwin a century back, illuminates our existence in a simple but almost overwhelming way."⁴

Simpson held a similar vision of human progress, but tempered it with a paleontologist's awareness of evolutionary regress. "I don't think that evolution is supremely important because it is my specialty," Simpson was fond of saying; "it is my specialty because I think it is supremely important." Understanding the evolutionary process empowers humans to control it, he explained, and by doing so they could avoid the fate of such previously dominant species as the dinosaurs. "A world in which man must rely on himself, in which he is not the darling of the gods but only another, albeit extraordinary, aspect of nature, is by no means congenial to the immature," Simpson observed around the centenary of Darwinism. "That is plainly a major reason why even now ... most people have not really entered the world into which Darwin led—alas!—only a minority of us." Now everyone should follow, he urged, and teaching the "fact of evolution" in all American public schools would serve as a vital step in that direction.[5] Having renounced the evangelical Protestantism of his childhood, Simpson easily slipped into the role of an evangelist for Darwinism.

Huxley and Simpson saw belief in God as a remnant of a prior stage in human psychosocial evolution. Like the human appendix, it no longer served an adaptive purpose and instead could cause harm.[6] The proclamation of this view by two prominent architects of the modern synthesis aroused opposition in America, where most people believe in an omnipotent God. Yet many key figures behind the modern synthesis disagreed with Huxley and Simpson about religion. Fisher clung to his Anglican heritage, for example, while Wright tilted toward Protestant process theology and Haldane warmed to Hinduism. Dobzhansky remained a professing Orthodox Christian and, during the 1950s, embraced the efforts of French Jesuit paleontologist Pierre Teilhard de Chardin to see nature evolving toward a final "Omega Point" of divine union in Christ. "Evolution (cosmic + biological + human) is going towards something, we hope some City of God," Dobzhansky wrote to a fellow Christian in 1961. "This belief is not imposed on us by our scientific discoveries, but if we wish ... we may see in nature manifestations of the Omega."[7] For his part, Lack converted from agnosticism to evangelical Protestantism in the very year he published *Darwin's Finches*. Like many mainstream Catholics and Protestants, he accepted evolution to a point, but believed that God created human souls. "Science," Lack wrote, "has not accounted for morality, truth, beauty, individual responsibility or self-awareness, and many people hold that, from its nature, it can never do so."[8] In practice, acceptance of the modern synthesis coexisted with all manner of religious faith.

———

Perhaps because of his conservative Christian upbringing at the dawn of the anti-evolution crusade, Simpson recognized the depth of popular opposition in America to the theory of evolution. Public-opinion polls suggested that nearly half of all Americans believed that God specially created humans within a biblical time frame, while most of the rest thought that God guided evolution. Less than one in ten respondents saw evolution as a purely natural process in line with the modern synthesis.[9] Simpson hoped that education could change this. "If a sect does officially insist that its structure of belief demands that evolution be false, then no compromise is possible," Simpson told schoolteachers in 1961. "Fortunately, the great majority of religious people in America belong to sects that are more flexible on this point."[10]

Simpson's comments ignored ongoing developments in American religion. An increasing percentage of religious people in America belonged to denominations or independent churches that displayed less (rather than more) flexibility on the interpretation of Scripture than ever before. If anything, the split between religious and scientific views on origins widened during Simpson's life. Various factors contributed to this peculiar development.

Theologically conservative churches in America grew faster than their mainline counterparts during the mid–twentieth century, and thereafter kept growing while many others began to contract. Among large Protestant denominations, the conservative Southern Baptist Convention and Lutheran Church–Missouri Synod outpaced all others in both percentage and absolute terms. Among small denominations, membership in fundamentalist and Pentecostal groups soared. The Mormons, Assemblies of God, and Seventh-day Adventists topped the chart in growth rates, with all three friendly to creationism.[11] Some of this shift came from parishioners moving from mainline Protestant denominations that they viewed as becoming too liberal, to churches professing strict biblical orthodoxy. Some came from an overall decline in church membership, which hit mainline denominations hardest.[12] Further, the South, where conservative churches dominated society and politics, gained influence during the period. More Americans moved south and Southern ways spread nationally, with the Southern Baptist Convention reaping a harvest from both trends.[13] As churches committed to biblical orthodoxy grew, they developed their own colleges, schools, publishing houses, broadcast media, and evangelistic associations—all poised to carry a creationist message. Finally, the secularization of elite society (which also emptied European churches) so sapped the vitality of mainstream American Protestantism that it could not counter fundamentalism as effectively as it once did. In the 1920s, prominent scientists from liberal churches took the lead in opposing the anti-evolution crusade. By the 1960s, that role passed largely to secular scientists, who were far less able to reach Christian audiences.

Although there were always Christians who opposed Darwinism, except during the anti-evolution crusade, fundamentalists tended to keep their concerns within the church—until the 1960s. The appearance in 1963 of the evolutionary BSCS high school biology textbooks changed the dynamics. These books, carrying the imprimatur of federal support, quickly became standard issue in up to half of the secondary schools in the United States, triggering an unexpected backlash. Protests by fundamentalists erupted in many places. Citing statements by Huxley and Simpson on science and religion, some argued that if teachers present atheistic evolution in biology classes then they should teach biblical creationism, as well. During the mid-1960s, in the first bills of their kind, two Arizona legislators introduced measures requiring that public schools give "equal time and emphasis to the presentation of the doctrine of divine creation where such schools...teach the theory of evolution."[14] In 1973, Tennessee (which had re-pealed its anti-evolution law six years earlier) became the first state to pass such legislation. Although protests rolled back the number of districts adopting BSCS texts, constitutional limits on religious instruction in public schools doomed the new Tennessee law and other efforts to teach the Bible in science classes.[15]

———

Diverging mid-twentieth-century developments in American science and religion drove a widening wedge between evolutionary biology and conservative Christianity. On the one hand, due to its materialistic, Darwinian orientation, the modern synthesis was less readily compatible with spiritual belief than earlier Lamarckian and theistic theories of evolution. Indeed, where many prominent American evolutionists once sought to reconcile their science with Christianity, by the second half of the twentieth century most neo-Darwinian biologists either repudiated or dismissed the effort. "I am a Christian," Dobzhansky observed in 1961, but Huxley's "is very much a majority opinion among at least the natural scientists."[16] On the other hand, the growing dominance of theologically conservative churches tilted the balance of American Protestantism toward the Right just as conservative Protestant opinion was hardening its interpretation of Genesis.

This hardening affected fundamentalist views on the age of the earth. Ever since Cuvier extended the time frame of geologic history, Christians have debated the meaning of passages in Genesis describing a six-day creation and a global flood survived by Noah. In the 1800s, many conservative

Christians equated the six days of creation with geologic ages and accepted the scientific view that the earth was very old. The Noachian flood receded in geologic significance. William Jennings Bryan still held this view in the 1920s, and testified to that effect on the witness stand at the Scopes trial.[17] Early in the 1900s, the enormously popular *Scofield Reference Bible* invoked a supposed gap in the Genesis chronology to leave time for innumerable geologic ages between the initial creation of "the heaven and the earth" and the subsequent creation of modern life forms.[18] The earth remained very old, but Genesis regained literal significance. At one time, the largely self-taught Seventh-day Adventist science teacher George McCready Price stood out among prominent American anti-evolutionists for his insistence on a truly literal reading of Genesis. In a deluge of publications over a thirty-five-year period ending with his retirement in 1938, Price argued for a six-day creation within a biblical time frame and a single catastrophic flood that shaped the earth's features and deposited the fossil record. He called his theory "flood geology." At the Scopes trial, Clarence Darrow dismissed Price as "the only human being in the world ... that signs his name as a geologist that believes" such things.[19] Through persistence, however, Price gained a following among fundamentalists. With each succeeding step—from day/age theology through the gap theory to flood geology—theologically conservative Christians hardened their interpretation of Genesis.

Even among fundamentalists, the appeal of flood geology was limited by Price's ties to the Seventh-day Adventist Church (which some Christians deemed heretical for the authority that it accorded the views of its founder-prophetess Ellen G. White, including her ecstatic vision of a six-day creation) and by his utter lack of any scientific credentials. This changed when the doctrine was reworked and reissued in 1961 by a Southern Baptist hydraulics engineer, Henry M.

Morris, and a Grace Brethren theologian, John C. Whitcomb, Jr. Each held a doctoral degree in his field. Both believed in the plenary verbal inspiration of Scripture. For them, every word in the Bible was from God and (as Morris once wrote) Christians should either "believe God's word all the way, or not at all."[20] Their 1961 collaborative work, *The Genesis Flood*, quickly became a fundamentalist classic. In it, they maintained that when Genesis says God created the universe in six days, it must mean six twenty-four-hour days; when it says that God created humans and all animal kinds on the sixth day, then dinosaurs must have lived alongside early humans; and when it gives a genealogy of Noah's descendants, believers can use it to date the flood at between five and seven thousand years ago. Most crucially, Morris and Whitcomb stressed biblical passages saying that God specially created each kind of life. Although this left room for the micro-evolution of related species, they asserted that it precluded the evolution of any one distinct type of living thing from another.[21]

With the publication of *Genesis Flood*, Morris became a popular speaker at fundamentalist churches, colleges, conferences, and camps across the country. More books and many articles followed. In 1970, he created what became the Institute for Creation Research (ICR), at Christian Heritage College in San Diego. The college was founded and led by Tim LaHaye, whose best-selling *Left Behind* series of novels would later do for biblical end-times prophecy what Morris's books did for biblical accounts of origins: make them come alive and be taken literally by millions of readers. Together, Morris and LaHaye covered the Bible from Genesis to Revelation.

Morris and his team of ICR scientists and educators opened a second front against the theory of evolution. Fundamentalists no longer merely denounced Darwinism as false; they offered a scientific-sounding alternative of their

Genesis Flood authors Henry M. Morris and John C. Whitcomb, Jr., 1984.

own, which they called either "scientific creationism" (as distinct from religious creationism) or "creation science" (as opposed to evolution science). "There seems to be no possible way to avoid the conclusion that, if the Bible and Christianity are true at all, the geological ages must be rejected altogether," a 1974 ICR publication stated. "The vast complex of godless movements spawned by the pervasive and powerful system of evolutionary uniformitarianism can only be turned back if their foundations can be destroyed, and that requires the re-establishment of special creation on a Biblical and scientific basis."[22] For some theologically conservative Christians intent on evangelism, the result was electric. "Within a decade or two the tireless proselytizers for scientific creationism had virtually coopted the generic creationist label for their hyperliteralist views, which only a half century earlier

had languished on the margins of American Fundamentalism," science historian Ronald Numbers notes.[23]

———

Not all American evangelicals, fundamentalists, and Pentecostals followed the ICR line, of course, but enough did to stir
up American education and politics. ICR promoted creation
science throughout the United States and elsewhere in books,
pamphlets, tapes, videos, lectures, and debates. Its biology
textbooks dominated the Christian and home-school markets.
During the mid-1970s, ICR sought to enter the public-school
market with a sanitized version of its text that omitted direct
references to God and the Bible. The use of this and other
creationist books, pushed by some fundamentalist parents
and groups but resisted in many places by science educators
and civil-liberties organizations, exacerbated the dispute over
teaching about origins in biology classes.

The constitutional case for including creation science in
American public schools (where religious instruction is
barred) rested on the claim that it serves a secular educational
purpose. Morris frankly admitted that teaching creation science promotes belief in a Creator. Indeed, he once called his
work "the *cutting edge* of the Gospel, the sharp wedge of foundational truth, in the great battle for the eternal souls of men
and women for whom Christ died."[24] Yet Morris and his followers maintained that the religious effect of such teaching
represents an incidental result of telling students the scientific truth about origins. And teaching the theory of evolution
also promotes a religious or philosophical worldview, they
added. In a book coauthored by his son John, Morris damned
belief in evolution as "the pseudo-scientific justification for
almost every deadly philosophy and every evil practice
known to man," including Social Darwinism, modern racism,
Nazism, Marxism, and the promotion of abortion, homosex-

uality, and illicit drug use.[25] Assuming that creation science is as scientific as the theory of evolution and that the theory of evolution is as much a philosophic viewpoint as creation science, the ICR asserted in a 1979 model school-board resolution that teaching both theories in public schools "would not violate the Constitution's prohibition against establishment of religion because it would involve presentation of the scientific evidences for each theory rather than any religious doctrine."[26]

In effect, the case for teaching creation science turned on the theory's scientific status. Virtually no secular scientists accepted the doctrines of creation science; but that did not deter creation scientists from advancing scientific arguments for their position. As presented in the ICR's basic biology textbook, these arguments are largely negative. No one can prove either creation or evolution by scientific observation or experiment because they occurred in the past or happen too slowly, it stated, yet one of these "two models" must account for life's diversity. The text then presented familiar objections to the theory of evolution: gaps in the fossil record; the improbability of random genetic mutations leading anywhere; the thermodynamic law against increasing order in physical systems; and the rational orderliness of nature. If these factors say no to evolution, it asserted, then they must say yes to creation. The textbook supplemented these points with controversial paleontologic evidence of early humans living among dinosaurs and a discussion of how one catastrophic flood could account for the earth's features and the fossil record.[27] Coming from experts with advanced degrees in science or engineering, and touted by prominent fundamentalist ministers, such arguments persuaded any number of theologically conservative Christians. By the 1980s, polling data suggested that most Americans favored teaching creation science alongside the theory of evolution in public schools.[28] The political impact of such opinions stunned mainstream biologists, who had generally ignored or ridiculed creation science.

The creationists' call for "balanced treatment" in teaching origins came at an opportune time for it to resonate loudly. Led by fundamentalist minister Jerry Falwell, evangelical theologian Francis Schaeffer, and Pentecostal televangelist Pat Robertson, conservative Christians became increasingly active in Republican Party politics during the years leading up to the 1980 elections, giving birth to the so-called "Religious Right" that has figured in American campaigns ever since. Legalizing school prayer and outlawing abortion stood at the top of the movement's agenda, but creationism in the classroom was part of its longer wish list. Republican presidential nominee Ronald Reagan signed on in 1980 when, in a speech to evangelical ministers, he characterized evolution as a "theory only" and endorsed the concept of balancing its teaching with creationist instruction.[29] He won his race with broad coattails, especially in the once-Democratic South. With support from the Religious Right, conservative Republicans upset Democrats in gubernatorial races in Louisiana and Arkansas that year. In 1981, those states became the first to enact statutes expressly requiring "balanced treatment for creation-science and evolution-science."[30] Upon signing his state's law, the new Arkansas governor stated, "If we're going to teach evolution in the public school system, why not teach scientific creationism? Both of them are theories."[31]

Not everyone viewed these laws as benign efforts to accommodate a diversity of viewpoints in the classroom. Some saw them as flagrant violations of U.S. Supreme Court rulings against promoting sectarian beliefs in public schools. If left unchallenged, they feared that other states or school districts might impose similar mandates. At ninety-three, ACLU founding executive director Roger Baldwin, who initiated the assault on the 1925 Tennessee anti-evolution law that culminated in the Scopes trial, was still active in 1981. "It is a strange feeling," he said. "Here's where I came in, and here's where the ACLU goes out to another battle to defend the same principle of freedom."[32] The ACLU filed lawsuits in federal courts to have both statutes declared unconstitutional. More than a hundred scientific, religious, or educational groups joined as plaintiffs or filed supportive legal briefs.

The ACLU moved first against the Arkansas law, which it viewed as more vulnerable than the Louisiana act. Sponsors had patterned both statutes after the ICR's model school-board resolution, but only the Arkansas law expressly defined "creation-science" to include scientific evidence for such Bible-based concepts as "separate ancestry for man and apes," a "worldwide flood," and the "relatively recent inception of the earth and living kinds."[33] This definition gave a height-ened sectarian taint to the statute that its opponents exploited in court. After hearing two weeks of testimony from experts in science and religion, trial judge William Overton con-cluded that creation science, as defined in the law, "is inspired by the Book of Genesis" and "has no scientific merit or edu-cational value as science." As such, he ruled, "the specific pur-pose" and "the *only* real effect of [the] Act is the advancement of religion." This made it unconstitutional.[34] In the wake of Overton's ruling, the judge reviewing Louisiana's act saw no need for a trial. "Because it promotes the beliefs of some the-istic sects to the detriment of others," he ruled in response to a pretrial motion, "the statute violates the fundamental First Amendment principle that a state must be neutral in its treat-ment of religion."[35] It was his ruling that creationists opted to appeal—all the way to the Supreme Court.

It took six years for the Louisiana Balanced Treatment Act to wind its way from the statehouse in Baton Rouge to the Supreme Court in Washington. After four separate rulings against the statute in lower courts, the Supreme Court had the final say. Justice William Brennan delivered its opinion in 1987. "The legislative history documents that the Act's primary purpose was to change the science curriculum of public schools in order to provide persuasive advantage to a particular religious doctrine that rejects the factual basis of evolution in its entirety," he concluded. This violates the Establishment Clause.[36] In a concurring opinion, Justice Louis Powell linked creation science to the Institute for Creation Research. Noting that "the Institute was established to address the 'urgent need for our nation to return to belief in a personal, omnipotent Creator,'" he noted, "it is clear that religious belief is the Balanced Treatment Act's 'reason for existence.'"[37] The Supreme Court's decision effectively barred the teaching of creation science in American public schools, but it did little to wean theologically conservative Christians from creation science. If anything, it accelerated the growth of Christian schools, where an ever-larger number of students learn their biology from ICR textbooks. The culture wars over origins scarcely paused.

—

Within a decade after the Supreme Court had spoken, members of a loosely organized coalition mostly composed of evangelical Christians were back before school boards, at statehouses, and in courts seeking authority for biology teachers to tell public-school students why the theory of evolution arouses controversy in science and society. Led by University of California law professor Phillip Johnson, a dynamic adult convert to evangelical Christianity, partisans for this cause demanded a place in the science curriculum for the concept that nature displays traces of intelligent design. Mainstream biologists countered that there is neither any controversy in science over whether species evolve nor any role for an intelligent designer in the evolutionary process. Scientists study natural causes, they maintained, not supernatural ones.[38] With lawyerly acumen, Johnson denounced such reasoning as circular. "We define *science* as the pursuit of materialist alternatives. Now what kind of answers do we come up with?" he asked. "By gosh, we come up with materialist answers!"[39] The modern synthesis may be the best naturalistic answer to the origin of species, Johnson pleaded, but still wrong. This mattered to him (a nonscientist) because he feared that naturalism in science fed nihilism in society. If instead we concede that nonnatural forces could shape the natural world, he argued, then the abrupt appearance of species in the fossil record and the intricate complexity of natural systems should support intelligent design over Darwinian evolution—and teachers should present both alternatives in public schools.[40]

By 2000, books by Johnson condemning naturalism in all its forms had become best sellers within the conservative Christian community and he had attracted a core following within academia. Biochemist Michael J. Behe and philosopher William A. Dembski stood near the core's center. Unlike Johnson, Behe did not deny the evolutionary concept of common descent, but he did assert that some biochemical processes (such as the cascade of multiple proteins required for blood clotting) are too irreducibly complex to have originated in the step-by-step fashion envisioned by the modern synthesis.[41] Recalling, in its way, the nineteenth-century claim that the eye could not have evolved piecemeal because it only functions as a whole, Behe maintained that an intelligence must have designed certain functional systems basic to life. This is an old argument, but Behe revived it with modern examples in his 1996 book, *Darwin's Black Box*. Darwinists replied with popular books of their own, giving evolutionary explanations for complex biological processes (including blood clotting).[42] Seeking to break the stalemate in God's favor, Dembski invoked probability filters (of the type used to sift radio signals from outer space for messages sent by intelligent beings) to suggest that life's complexity is more likely the product of design than chance, but probabilistic arguments never carry much weight in Darwinian circles. Even one in a zillion can happen, some Darwinists stress; virtually alone in a vast and ancient universe, we could be it. After hearing Dembski speak in 2000, for example, biologist Kenneth R. Miller denounced the idea of intelligent design as an "imposter masquerading as a scientific theory."[43]

The new millennium began with the American public as divided as ever over the origin of species. In their federally mandated education standards, most states required that public secondary schools teach the theory of evolution in biology courses, but some individual school districts or teachers still downplayed the concept. Alabama offset its evolution-teaching mandate with a prefatory comment in its science standards: "Explanations of the origin of life and major groups of plants and animals, including humans, shall be treated as theory and not as fact."[44] In 1998, the elected Kansas board of education voted to delete any requirement to teach about macro-evolution from its science standards, but reversed itself two years later after several conservatives lost their seats on the board in the ensuing election. And in 2001, through the efforts of Republican senator Rick Santorum, Congress included language drafted by Johnson in the conference report for a federal education bill. "Where topics are taught that may generate controversy (such as biological evolution)," the report noted, "the curriculum should help students to understand the full range of scientific views that exist, why such topics may generate controversy, and how scientific discoveries can profoundly affect society."[45] These perceived social effects of the scientific theory ensured that the popular controversy over creation and evolution would continue. Indeed, late-twentieth-century developments in evolutionary biology made them loom larger than they had since the heyday of Social Darwinism.

CHAPTER 12

POSTMODERN
DEVELOPMENTS

Francis Crick winged into the Eagle, a pub popular with researchers at Cambridge University's nearby Cavendish Laboratory, boasting to one and all, "We have found the secret of life." It was early in 1953, and the "we" referred to thirty-six-year-old British biophysicist Crick and twenty-four-year-old American biochemist James D. Watson, then working at the Cavendish on a postdoctoral fellowship. Watson reported feeling "slightly queasy" at Crick's boast, but over the next few decades, many biologists came to see it as fully justified.[1] In one of the great "eureka" experiences of modern science, Watson and Crick had discovered the gene's double-helical structure by brilliantly and rapidly combining the findings of others with insights of their own. This surprisingly simple, highly elegant structure shed new light on the mechanics of evolution by suggesting how genetic reproduction, inheritance, and variation operated at the molecular level.

Although the gene stood at the heart of the modern synthesis, it was a black box prior to 1950. Until then, many scientists envisioned the gene as a complex assemblage of proteins that would take decades to decipher. Yet a growing body of evidence suggested that a much simpler macromolecule, deoxyribonucleic acid (or DNA), carries hereditary information. Watson and Crick followed the latter trail, and it led them to glory. They found that DNA is structured somewhat like a twisted railroad track with sturdy rails along its outer edges and a sequence of connecting ties, each composed of one of two different pairings of four base molecules commonly identified by their initials: A, T, G, and C. If DNA

splits lengthwise, then each half replicates the whole by at-
tracting new pairs for its remaining bases from the cell's organic soup, A to T and G to C. The macromolecule carries
genetic information in the sequence of its base molecules,
which serve as a template for forming ribonucleic acid (or
RNA) and, in turn, proteins. Information flows only one way
in this mechanism—from the DNA to the proteins that con-
struct the organism; never from the organism back to the
DNA. The result nicely matches the neo-Darwinian principle that inborn hereditary information guides individual development without any gene-altering feedback from the
environment. In these and other respects, DNA structure
provides a serviceable molecular foundation for evolution to
proceed in a manner fitting the modern synthesis. Both concepts are starkly materialistic and functionally reductionist.
Still, tensions developed between molecular biologists and
neo-Darwinian evolutionists.

Watson and Crick did not work out all the implications of
DNA structure themselves. In their initial 1953 papers, they
simply noted that it "immediately suggests a possible copying
mechanism for the genetic material," which they spelled out in
some detail, and that "spontaneous mutations may be due to a
base occasionally occurring in one of its less likely tautomeric
forms."[2] These and other implications inspired a generation of
scientists to pursue molecular biology. Traditional ways of
studying evolution suddenly seemed terribly old-fashioned.
"For those not studying biology at the time in the early 1950s,
it is hard to imagine the impact the discovery of the structure
of DNA had on our perception of how the world works,"
zoologist Edward O. Wilson later recalled. "If heredity can be
reduced to a chain of four molecular letters—granted, billions
of such letters to prescribe a whole organism—would it not
also be possible to reduce and accelerate the analysis of
ecosystems and complex animal behavior?"[3]

Watson and Wilson, who became two of the most influential scientists of the late twentieth century, both joined Harvard's biology department as assistant professors in 1956. Watson led the shock troops for the newer forms of molecular biology. Wilson, who studied ants, upheld the older naturalist tradition associated with Ernst Mayr and the modern synthesis. The two young biologists rarely spoke to each other during their years together at Harvard. Their department, in a microcosm of larger developments within the profession, eventually split into separate ones for molecular and evolutionary biology. Wilson later described Watson as "the most unpleasant human being that I had ever met."[4] Their relationship worsened after Wilson (whom Watson regarded as a mere bug collector) received tenure before Watson. Mayr, Wilson, and others in their camp viewed molecular biology as too narrow and limited to comprehend all aspects of the evolutionary process. To them, organisms and ecosystems still mattered. For his part, in the best-selling book *The Double Helix,* Watson dismissed most zoologists and botanists of the 1950s as "a muddled lot [who] wasted their efforts in useless polemics about the origin of life." Geneticists of the era fared little better in Watson's account. "You would have thought that with all their talk about genes they should worry about what they were," he wrote. "All that most of them wanted out of life was to set their students onto uninterpretable details of chromosome behavior or to give elegantly phrased, fuzzy-minded speculations over the wireless on topics like the role of the geneticist in this transitional age of changing values."[5] Yet over time, and from their separate departmental homes, evolutionary and molecular biology grew to complement and reinforce each other.

At the elemental level, the discovery that all species (even the most primitive unicellular ones) shared a common genetic code suggested that they have a common ancestry. In turn, the comparative study of DNA from various organisms elucidated their evolutionary relationships. In one spectacular example of this from the 1960s, Dobzhansky's protégé Richard C. Lewontin used a technique called "gel eletrophoresis" to measure genetic variation among individuals of the same species. This analysis tested his mentor's hypothesis that enough latent variability exists in recessive alleles to feed the evolutionary process in response to changed environmental conditions without added mutations. Lewontin found what he was looking for—indeed, he found so much genetic variability within species that much of it must have little or no effect on individuals.[6] Taking the position that all the variation may be meaningless, diehard opponents of Dobzhansky's hypothesis clung to the classical view (historically associated with Thomas Hunt Morgan and Hermann Muller) that mutations feed evolution.

Further, Watson's success in reducing much of biology to molecules inspired even his adversaries. "He and other molecular biologists conveyed to his generation a new faith in the reductionist method of the natural sciences," Wilson later noted. "A triumph of naturalism, it was part of the motivation for my own attempt in the 1970s to bring biology into the social sciences through a systematization of the new discipline of sociobiology."[7] Watson came to appreciate Wilson's work in sociobiology—and it helped mend their relationship in the years after Watson left Harvard to run the Cold Spring Harbor genetics laboratory, a traditional center for research in applied human genetics, dating from the heyday of eugenics. Yet Wilson never took a molecular approach to sociobiology. Indeed, his initial foray into the field grew out of his interest in how insect colonies function, which he pursued in opposition to the molecular focus of biologists who followed in Watson's wake. Early on, for example, Wilson explored the impact of population size and density on the caste system of ant colonies and on aggressive behavior by various types of "social animals" (or animals that live in groups). Molecules alone could not explain these developments, he argued.

Instead, the fundamental breakthrough to sociobiology came in a brilliant two-part 1964 article by William D. Hamilton, a British graduate student who also rejected the molecular approach to biology. As a self-proclaimed disciple of Ronald Fisher, Hamilton focused his powerful intellect on interactions among genes rather than on the gene's molecular composition. Here, he believed, lay the true secret of life. While still a college student at Cambridge during the 1950s, Hamilton later wrote, "I was convinced that none of the DNA stuff was going to help me understand the puzzles raised by my reading of Fisher and Haldane or to fill the gaps they had left. Their Mendelian approach had certainly not been outdated by any new findings."[8] Ultimately, Hamilton exceeded even Fisher in seeing evolution (as he put it in his 1964 article) "from a gene's point of view," and he inspired Wilson and other sociobiologists to do so, as well.[9]

The origins of altruism stood out as the most prominent evolutionary puzzle unsolved by the synthetic theorists. They had melded Darwinism and Mendelism without accounting for self-sacrifice. This left a glaring gap. Critics of selection theory had long pointed to altruistic behavior (particularly by humans) as evidence that a Darwinian struggle for existence could not explain all aspects of life. If nature selects solely for traits leading to an individual's survival or reproductive success, as classical Darwinism suggested, then self-sacrifice (except to aid one's own descendants) must have a supernatural source. Such thinking led Darwinists from Alfred Russel Wallace through David Lack to reserve space for the spiritual within their science. Darwin held out for a naturalistic explanation for altruism, but never devised a wholly satisfactory one. Although he otherwise thought that selection acted among individuals, here Darwin bowed to group selection.

Altruism aids the group at the expense of the individual, he reasoned, such as when a bird warns the flock of approaching predators by a self-endangering cry or when childless soldiers die for their country. Castes of sterile worker ants, wasps, and bees present even more dramatic examples of self-sacrificing social behavior. Inasmuch as such traits promote the group's survival, Darwin noted, perhaps nature selects groups that possess them. Unless members of the group jointly learn and pass along such traits in a Lamarckian manner, however, those traits should fail because single individuals displaying them would tend to die earlier than others. In short, as a randomly generated inborn trait, altruism should not persist. Such thinking led Darwin to admit an element of Lamarckism into his system—but the modern synthesis rejected this solution.

Hamilton proposed a purely Darwinist account for altruism by shifting the level of selection to the gene. Take social insects, he proposed. Because of the peculiar way they reproduce, female ants, wasps, and bees share more genes in common with their sisters (75 percent) than with their own children (50 percent) or brothers (25 percent). Thus, from the gene's point of view, female ants achieve greater reproductive success by aiding their sisters than their offspring. Fitting the model, all sterile worker ants are female and, as Hamilton noted, "working by males seems to be unknown in the group."[10] Similarly but less spectacularly, any animal (including a human) that shares genes with its collateral kin (as well as with its lineage) can maximize the survival of those genes by sacrificing for its relatives provided the number of genes held in common by those relatives exceed those lost by its sacrifice. In his 1964 article, Hamilton worked out the algebra of such so-called "kin selection" to show that, at least in theory, it could account for apparently altruistic behavior in terms of selfish genes. "In the world of our model organisms," he concluded, "we expect to find that no one is prepared to

sacrifice his life for a single person but that everyone will sacrifice it when he can thereby save more than two brothers, or
four half brothers, or eight first cousins. . . ."[11] Genetic tendencies in this direction should survive and spread. In this
manner, a starkly naturalistic struggle for existence could lie
at the heart of seemingly selfless acts of individual love. With
this conceptual breakthrough, the way seemed open to find
biologic bases for all manners of human and other-animal behavior.

———

Wilson first read Hamilton's paper in 1965, on a train trip
from Boston to Miami. By his own account, Wilson left New
England with a casual interest in Hamilton's ideas, became
increasingly "frustrated and angry" about them during the
journey, but arrived in Florida fully convinced. "Because I
modestly thought of myself as the world authority on social
insects," Wilson related, "I also thought it unlikely that anyone else could explain their origin, certainly not in one clean
stroke," but Hamilton had.[12] Like Hamilton, Wilson believed
that whatever explained the behavior of social insects also
shed light on the behavior of other social animals, including
humans. A core group of evolutionary biologists agreed. An
outburst of research ensued, testing and extending Hamilton's insight and other sociobiologic theories (collectively
called "evolutionary psychology" when applied to humans).
These efforts sought to explain behavior in terms of its impact on the survival and reproductive success of genes and individuals. Models, metaphors, and concepts proliferated,
such as reciprocal altruism (where organisms evolve to help
one another survive), evolutionarily stable strategies (where a
balance of behaviors within a population serves individual
interests), and an arms race (where predator and prey evolve
in response to one another's developments). By factoring in
circumstances impacting populations, sociobiologists used

such theories to predict and explain all manners of animal behavior.

Wilson pulled many of these threads together in his 1975 survey, *Sociobiology: The New Synthesis*. Because of its emphasis on adaptations aiding reproduction, the book featured biologic accounts of gender-based behaviors. Males naturally tend to spread their ample sperm (including through multiple mates), Wilson suggested, while females tend to conserve their scarce eggs (such as by investing heavily in mate-selection and child-rearing). Indeed, some sociobiologists accounted for aggressive behavior by young males as a genetic holdover from a time when it carried reproductive benefit. Among chimpanzees, for example, the most sexually aggressive males produce the most offspring. To some readers, such explanations sounded like justifications for traditional gender roles and sociobiology seemed to endorse the social status quo (or worse)—especially given Wilson's insistence that humans disregard nature at their peril.

Characteristically, the chapter of *Sociobiology* dealing with human behavior opened with a deliberately provocative challenge. "Let us now consider man in the free spirit of natural history," Wilson wrote. "In this macroscopic view the humanities and social sciences shrink to specialized branches of biology; history, biography, and fiction are the research protocols of human ethology; and anthropology and sociology together constitute the sociobiology of a single primate species." Ethics, he suggested, should "be removed temporarily from the hands of the philosophers and biologicized."[13] These were bold claims for a discipline supposedly chastened by the excesses of Social Darwinism, eugenics, and Nazi race theory. Wilson did not assert that nature alone shaped human behavior, but he clearly wanted to push the pendulum back from the extreme nurture position taken by most mid-twentieth-century social scientists. They envisioned human

Edward O. Wilson in his office, 2003.

culture as infinitely malleable for good or ill while he declared that "the gene holds culture on a leash."[14]

The reaction came swiftly. Steeped in environmentalism,
many social scientists and humanists hotly contested Wilson's
claims, with protesters once going so far as to pour ice water
on him during an academic address. Following the lead of
T. H. Huxley, who in 1893 called it a "fallacy" to base ethics
on evolution, many biologists had conceded the study of

human behavior to the social scientists.[15] "Culture is acquired, not transmitted through genes," Dobzhansky assured anthropologists in 1963.[16] Beginning in 1975, Dobzhansky's former student and Wilson's colleague, Richard Lewontin, led the scientific assault on sociobiology. Another distinguished Harvard evolutionist with a popular following, paleontologist Stephen Jay Gould, joined the attack. Both acknowledged that genetic determinism offended their Marxist ideology, just as it had offended Dobzhansky's Christian beliefs, but they focused their criticisms on Wilson's science. Likening sociobiologic explanations of human behavior to Rudyard Kipling's "just so" fables of how primitive peoples account for animal origins, Lewontin and Gould damned sociobiology as scientifically flawed and socially dangerous. "Wilson joins the long parade of biological determinists whose work has served to buttress the institutions of their society by exonerating them from responsibility for social problems," they wrote in a 1975 critique of *Sociobiology,* cosigned by a dozen other Boston-area academics.[17]

Wilson replied with an expanded defense of human sociobiology in his award-winning 1978 book, *On Human Nature.* The sparring continued into the twenty-first century, with Wilson gradually gaining allies among evolutionary biologists and a new generation of social scientists. "Overall," he boasted in the foreword to a commemorative twenty-fifth-anniversary edition of *Sociobiology,* "there is a tendency at the century's close to accept that *Homo sapiens* is an ascendent primate, and that biology matters."[18] Indeed, by offering edgy, materialistic explanations for human interactions, sociobiology and gene-centered evolutionism attracted a popular (as well as an academic) following. It fit the secular, consumer-oriented culture commonly associated with America in the 1980s. Although books by Wilson on the topic sold well, they were eclipsed on both sides of the Atlantic by those of a younger British evolutionist, Richard Dawkins, who as an Oxford student in the late 1960s had taken Hamilton as his "intellectual hero."[19]

In his intoxicating prose, Dawkins popularized Hamilton's vision of organisms (including humans) as elaborate apparatuses evolved to propagate their genes. "We are survival machines—robot vehicles blindly programmed to preserve the selfish molecules known as genes," he explained. "This is a truth which still fills me with astonishment."[20] The genes themselves cannot plan ahead or respond to their environment, Dawkins stressed. They simply reproduce themselves with occasional random mutations that may or may not assist their survival, and we ("climbing Mount Improbable" of Sewall Wright's adaptive landscape over some four billion years of trial-and-error organic evolution) are the result.[21] Dawkins found this view of life exhilarating. For him, it freed humans from the burden of purposeful design in nature, which he identified as "the most influential of the arguments for the existence of God." Unlike the controlling purposes of a designing God, Dawkins noted, "Natural selection, the blind, unconscious, automatic process which Darwin described, and which we now know is the explanation for existence and apparently purposeful form of all life, has no purpose in mind." By banishing the argument for God from design, he proclaimed, "Darwin made it possible to be an intellectually fulfilled atheist."[22]

Agreeing with Dawkins's naturalistic gospel in so far as it went, Wilson nevertheless hoped for something more from scientific materialism. "It presents the human mind with an alternative mythology that until now has always, point for point in zones of conflict, defeated traditional religion," he stated. "Its narrative form is epic: the evolution of the universe from the big bang . . . [to] life on earth."[23] Reared in the Bible Belt of Alabama by fundamentalist parents, Wilson maintained that people need to believe in something larger than themselves to justify the individual sacrifices that propagate genes through kin selection. Indeed, in an article coauthored by Darwinian philosopher Michael Ruse, Wilson described religion (or at least ethics based on religion) as "an illusion fobbed off on us by our genes to get us to cooperate."[24] As people shed spiritual belief in the light of scientific understanding, he believed that some other source of larger meaning must take its place. In his 1998 book *Consilience* and elsewhere, Wilson offered evolutionism as a new "sacred narrative" capable of enshrining essential ethical principles calculated to advance human development and preserve genetic diversity.[25] Although Wilson's vision of a naturalistic religion based on modern evolutionary thought stirred widespread comment, most scientists kept their professional distance. "He is a good and gentle man, generous to a fault, with a real concern for... biodiversity and ecological preservation," Ruse noted. "But I do not see that his fellow evolutionists have to follow him in making a religion out of their shared science."[26]

Of course, scientists can see their science as important without making it their religion. Certainly most evolutionary biologists view the theory of evolution as extremely important: That is why they study it. Hamilton, for example, like Fisher before him, entered the field in pursuit of the eugenicist's dream of enhancing the human stock. "I had come to Galton's ideas by my own parallel reasoning spurred by the common youthful wish to improve the world, and by reading Fisher," Hamilton wrote shortly before his death in 2000. "I much liked the notion that human-directed selection, whether to maintain standards or to speed the intellectual and physical progress of humanity, could be made more effective and more merciful than the obviously inefficient and cruel natural process."[27] Although this dream sustained his early studies of altruistic behavior, Hamilton's later work on

another major evolutionary puzzle left unsolved by Fisher—
the so-called "problem of sex"—dampened his enthusiasm
for state-controlled eugenics and deepened his appreciation
for the efficiency of natural selection.

—

Sexual (as opposed to asexual) reproduction comes at a huge
biologic cost to the individual. The birthing female transmits
only half of her individual genetic identity to her offspring.
The payoff comes in variation as the genetic material from
two parents passes on in new combinations. Under evolutionary models that rely exclusively on individual or gene-level
selection (rather than group selection or intelligent design),
synthetic theorists began questioning whether the payoff justified the cost. Among biologists taking up this puzzle, which
included such leading neo-Darwinists as George C. Williams
and John Maynard Smith, Hamilton offered perhaps the most
convincing (though far from universally accepted) reason
why sexual reproduction predominates among species that
produce few offspring. He called it the "parasite Red Queen
hypothesis," in reference to the character in *Through the
Looking-Glass* who had to run as fast as she could just to stay in
place—a name borrowed from biologist Leigh Van Valen's
more general "Red Queen" hypothesis of an evolutionary
arms race. It relied on the premise that sexual reproduction
produces more genetic variation than asexual reproduction.
Hamilton proposed that slow-reproducing organisms need
the added variation coming from sexual reproduction to stay
ahead of their rapidly reproducing asexual parasites.[28] "The
hosts' best defense may be based on genotypic diversity,
which, if recombined each generation, can present to the parasites what amounts to a continually moving target," he explained in a coauthored 1988 article.[29]

The parasite Red Queen hypothesis, with its emphasis on
random genotypic diversity, convinced Hamilton that any

controlled scheme of eugenic breeding would fail through a lack of genetic variation. Better let natural selection take its course, he concluded, whatever the pain.[30] Yet his youthful enthusiasm for planned reproduction merely gave way to later-life concerns about the dysgenic effects of advances in health care that allow people (such as diabetics, he noted) to survive and reproduce their deleterious genes. "I predict that in two generations the danger being done to the human genome by the anti- and postnatal life-saving efforts of modern medicine will be obvious to all," he warned in a posthumously published autobiographical account.[31] Hamilton simply could not stop worrying about humanity's future: Perhaps it was in his genes.

Just as Gould joined in leading the scientific opposition to sociobiology, he loudly protested any return to eugenic thinking. Gould saw the two as inexorably linked in a damnable hereditarian view of humanity. During the 1980s, he wrote popular books and articles exposing "the mismeasure of man" that bedeviled earlier efforts to devise eugenic standards for human reproduction.[32] Hamilton attributed Gould's position to a Marxist's misguided commitment to human equality, while Ruse viewed it as at least partially rooted in a Jew's memory of the Holocaust.[33] Indeed, Gould did not limit his assault to sociobiology and eugenics, but took on fundamental tenets of the modern synthesis that supported them. In doing so, he offered an alternative scientific view of how evolution operates that challenged genetic reductionism.

———

Gould complained that exclusive reliance on the natural selection of genes to account for evolution ignored factors that shape organisms. With Richard Lewontin in 1979, Gould argued that developmental constraints limit and channel adaptations. Certain features (such as the *Tyrannosaurus*'s reduced front legs) may serve no adaptive purpose, they suggested, but instead arise as a by-product of other adaptions (such as larger hind legs).[34] Over the years, Gould expounded the view that form does not inexorably follow function, and that much is left to chance in the lottery of life. "If a large ex-traterrestrial object—the ultimate random bolt from the blue—had not triggered the extinction of dinosaurs 65 mil-lion years ago, mammals would still be small creatures, con-fined to nooks and crannies of a dinosaur's world, and incapable of evolving the larger size that brains big enough for self-consciousness require," he commented in 1996.[35] Gould's view flew in the face of the progressivism implicit in much modern neo-Darwinian thinking. Given the enormous adaptive advantage of intelligence and cooperation, for exam-ple, both Hamilton and Wilson believed that the evolution of self-conscious, altruistic beings was far from accidental. Their depictions of kin-based and reciprocal altruism in various types of social animals supported their faith in evolutionary progress. "Since natural selection has invented both kinds of altruism numerous times," the like-minded science writer Robert Wright explained, "it is not too wild to suggest that this expansive sentiment was probable all along."[36]

Beginning in the 1970s, Gould worked with American Museum of Natural History paleontologist Niles Eldredge in formulating the theory of punctuated equilibria to account for the pattern of organic life preserved in sedimentary rock. "The oldest truth of paleontology proclaimed that the vast majority of species appear fully formed in the fossil record and do not change substantially during the long period of their later existence," Gould noted in an obvious reference to Cuvier's findings. "In other words, geologically abrupt ap-pearance followed by subsequent stability."[37] The modern synthesis, in contrast, presented evolution proceeding gradu-ally, by minute adaptations, without sharp divisions of individuals into enduring species. For generations, Darwinists confidently predicted that further research would smooth out the fossil record into the predicted pattern of gradual change, but it never happened. Increasingly, synthetic theorists extrapolated from their mathematical models and population studies to chart the course of evolution, while relegating fossils to museum exhibits calculated to impress the public that evolution happens. Gould and Eldredge resisted this role for their field. Instead, they took its findings seriously, and tried to explain them in evolutionary terms. To do so, they drew on Sewall Wright's classic (but unfashionable) notion of nonadaptive genetic drift as developed in Ernst Mayr's so-called "founder principle" of allopatric speciation.

Mayr had proposed that new species form when a small population becomes geographically isolated from the main group—such as a few mainland finches blown to the Galápagos Islands. The new or particularized environment, coupled with a greatly restricted gene pool, accelerates the evolutionary process and facilitates the formation of new species, Mayr suggested. Once established, though, the new species should become as stable as the old one. This process, Eldredge and Gould observed, should generate the pattern they found in the fossil record: long periods of equilibria or stasis in species, punctuated with the abrupt appearance of new ones. "If new species arise very rapidly in small, peripherally isolated local populations, then the great expectation of insensibly graded fossil sequences is a chimera," they wrote in 1972. "A new species does not arise from the slow transformation of all its forebears. Many breaks in the fossil record are real."[38]

By the time Eldredge and Gould proposed their theory, however, Mayr's founder principle played little part in mainstream evolutionary thought, which Eldredge characterized as having become ultra-Darwinian. "Ultra-Darwinians are really followers of Ronald Fisher," he asserted, "seeking to explain all evolutionary phenomena strictly in terms of natural selection acting on heritable variation within populations."[39] Like Wright's concept of genetic drift, synthetic theorists came to view the founder principle as insignificant in its evolutionary effect as compared with gene-based adaptations in large populations—such as the selection of black peppered moths in a darkened environment. Now Eldredge and Gould sought to revive it as a primary explanation for the patterns preserved in the fossil record and, in doing so, to give special significance to speciation within the overall evolutionary process. "Species represent a level of permanence that acts to conserve adaptive change far beyond the ephemeral capacities of local populations," Eldredge asserted.[40] Drifting further from the neo-Darwinian mainstream, Gould later proposed that nonadaptive developmental constraints and macromutations might also impact evolution, at least at higher levels of the process. There is something real about species, he stressed, that resists change without a jolt beyond that required for lower-level variation among individuals and within populations. Using language certain to inflame synthetic theorists who believed that evolution operated the same way at all levels, Gould differentiated between "microevolution" within species and "macroevolution" of higher-level taxa.[41]

Gould's ability to communicate his heresies to an educated audience through best-selling books and popular articles fed the public perception of a revolution within evolutionary thought, when in reality the rebellion was limited largely to one wing of paleontology. Even Gould soon retreated from his grander claims. For his part, Eldredge maintained that the "abrupt appearance" of any new species under the punctuated-equilibrium model would still take generations of gradual change—he estimated it as ranging "from five to fifty thousand years"—and did not involve macromutations.[42] Any plausible version of punctuated equilibria, Mayr insisted, is compatible with the modern synthesis. Nevertheless, purporting to speak for mainline neo-Darwinists generally, Maynard Smith dismissed Gould's view of evolution as "so confused as to be hardly worth bothering with."[43] Robert Wright added, "He stresses its flukier aspects—freak environmental catastrophes and the like—and downplays natural selection's power to design complex life forms. In fact, if you really pay attention to what he is saying, and accept it, you might start to wonder how evolution could have created anything as intricate as a human being." Gould's view thus contrasted sharply with that of Hamilton, Wilson, and other ultra-Darwinists who saw the creation of humans as all but inevitable in the naturalistic processes envisioned by the modern synthesis. Wright dismissively dubbed Gould as "the accidental creationist," even though Gould steadfastly proclaimed his commitment to evolutionary materialism.[44]

———

Of course, many evolutionists still dissent on the critical issue of materialism, particularly when it comes to the ascent of humans. Calling himself a "theistic evolutionist," for example, noted American geneticist Francis Collins, who directed the human genome project during the 1990s and beyond, followed the path trod by God-fearing Darwinists from Wallace to Lack in rejecting naturalistic explanations for altruism and other distinctly human traits. "Science," Collins wrote in 2002, "will certainly not shed any light on what it means to love someone, what it means to have a spiritual dimension to our existence, nor will it tell us much about the character of God."[45] Surveys from the 1990s suggested that Collins speaks for about 40 percent of the American people—and almost a like percentage of American scientists—when he posits God as somehow involved in the evolutionary process, at least to the extent of breathing a supernaturally created soul into a naturally evolved body.[46] For the record, the Roman Catholic

Church approves the latter position. "Rather than *the* theory of evolution, we should speak of *several* theories of evolution... materialistic, reductionist and spiritualist," Pope John Paul II declared in a 1996 message to the Pontifical Academy of Sciences. "Theories of evolution which, in accordance with the philosophies inspiring them, consider the mind as emerging from the forces of living matter, or as a mere epiphenomenon of this matter, are incompatible with the truth about man."[47]

Despite challenges, the naturalistic modern synthesis stands at the heart of current evolutionary science. Genetic mutations and recombinations, this view maintains, cause organisms to vary. The fittest of these various individuals survive to pass along their genes. Change builds incrementally, without discrete breaks. As synthetic botanists have asserted all along, hybrid crosses between nearly related species add to the gene flows that cause individuals to vary.[48] Recent research suggests that (much as happens in genetic engineering) viruses and bacteria can invade the cells of other organisms and implant their own genes or genes from other organisms into the host's DNA.[49] Again, variation can result. Natural hybridization and gene acquisition join mutation and recombination as the genetic fodder for natural selection to sift and winnow in evolving the diversity of life.

Two centuries of research in biology since Lamarck's day have enriched the scientific account of origins that Wilson now calls "epic." "It is indeed remarkable that this theory has been progressively accepted by researchers, following a series of discoveries in various fields of knowledge," John Paul II noted in his 1996 message. "The convergence, neither sought nor fabricated, of results of work that was conducted independently is in itself a significant argument in favor of this theory."[50] Indeed, its proponents describe it as a "fact," not a theory.[51] The evolutionary epic began with the appearance of self-replicating cells on earth some four billion years ago, scientists estimate. Multicellular life came more than three billion years later; hominids stood erect on the plains of Africa beginning about five million years ago; and the first of our species walked into history within the past hundred thousand years. "There is a grandeur in this view of life," Darwin wrote in the last sentence of *Origin of Species,* "that, whilst this planet has gone cycling on according to the fixed law of gravity, from so simple a beginning endless forms most beautiful and most wonderful have been, and are being, evolved."[52] With this, all manner of modern evolutionists would agree. Even if they cannot wholly accept Wilson's depiction of "a cause-and-effect continuum from physics to the social sciences, from this world to all other worlds in the visible universe, and backward through time to the beginning of the universe,"[53] they share the awe implicit in a well-known movie's title that Gould adapted in naming his most popular book. Accident or not, *It's a Wonderful Life.*[54]

NOTES

CHAPTER 1. BURSTING THE LIMITS OF TIME

1. Georges Cuvier, "Sur un nouveau rapprochement à établir entre les classes qui composent le règne animal," *Annales du Muséum d'Histoire Naturelle*, 4 (1812), 73.

2. Georges Cuvier, "Memoir on the Species of Elephants, Both Living and Fossil," in Martin J. S. Rudwick, *Georges Cuvier, Fossil Bones, and Geological Catastrophes: New Translations and Interpretations of the Primary Texts* (Chicago: University of Chicago Press, 1997), 24.

3. Georges Cuvier, "Preliminary Discourse," in Rudwick, *Georges Cuvier,* 185.

4. *Ibid.,* 183.

5. Stephen Jay Gould, "Wide Hats and Narrow Minds," *Natural History,* Feb. 1979, 35.

6. *Ibid.,* 35.

7. Georges Cuvier, "Memoir on the Almost Complete Skeleton of a Little Quadruped of the Opossum Genus, Found in Plaster Stone Near Paris," in Rudwick, *Georges Cuvier,* 72.

8. Georges Cuvier, *Rapport historique sur les progrès des sciences naturelles depuis 1789* (Paris: 1810; fasc. rpt. Brussels: Culture et Civilisation, 1968), 11.

9. Cuvier, "Memoir on the Species of Elephants," 19.

10. Georges Cuvier, "Extract from a memoir on an animal of which the bones are found in the plaster stone around Paris, and which appears no longer to exist alive today," in Rudwick, *Georges Cuvier,* 36.

11. Cuvier, "Preliminary Discourse," 217.

12. Georges Cuvier, "Extract from a work on the species of quadrupeds of which the bones have been found in the interior of the earth," in Rudwick, *Georges Cuvier,* 45–46.

13. Georges Cuvier, "Historical Report on the Progress of Geology since 1789," in Rudwick, *Georges Cuvier,* 124.

14. Cuvier, "Extract from a work on quadrupeds," 52.

15. Cuvier, "Preliminary Discourse," 189.

16. *Ibid.,* 226.

17. *Ibid.,* 229.

18. *Ibid.,* 190.

19. William Coleman, *Georges Cuvier: Zoologist* (Cambridge: Harvard University Press, 1964), 139.

20. *Ibid.,* 240, 248.

21. E.g., Georges Cuvier, *Essay on the Theory of the Earth with Geological Illustrations by Professor Jameson,* 5th ed. (Edinburgh: Blackwell, 1827), xvi–xvii.

CHAPTER 2. A GROWING SENSE OF PROGRESS

1. Charles Darwin, in Stephen Jay Gould, *Time's Arrow, Time's Cycle: Myth and Metaphor in the Discovery of Geological Time* (Cambridge: Harvard University Press, 1987), 99.

2. William Buckland, "Notice on the Megalosaurus or Great Fossil Lizard of Stonesfield," *Transactions of the Geological Society* [London], ser. 2, vol. 1 (1824), 390.

3. *Ibid.,* 390–91.

4. *Ibid.,* 391.

5. Georges Cuvier, in David A. E. Spalding, *Dinosaur Hunters* (Rocklin, CA: Prima, 1993), 21.

6. Gideon Algernon Mantell, "The Geological Age of Reptiles," *Edinburgh New Philosophical Journal,* 11 (1831), 181–85.

7. A much-quoted quip reported in Roger Hahn, "Laplace and the Mechanical Universe," in David C. Lindberg and Ronald L.

Numbers, *God and Nature: Historical Essays on the Encounter Between Christianity and Science* (Berkeley: University of California Press, 1986), 256.

8. William Buckland, *Geology and Mineralogy Considered with Reference to Natural Theology,* 1 (London: William Pickering, 1837), 8–9.

9. Gould, *Time's Arrow,* 99.

10. Adam Sedgwick to Louis Agassiz, 10 Apr. 1945, in Elizabeth Cary Agassiz, ed., *Louis Agassiz: His Life and Correspondence* (Boston: Houghton, Mifflin and Co., 1886), 384–85 [emphasis in original].

11. Adam Sedgwick, "Address to the Geological Society, Delivered on the Evening of the 18th of February, 1831," *Proceedings of the Geological Society of London,* 1 (1826–33), 306.

12. *Ibid.,* 383.

13. Adam Sedgwick, *A Discourse on the Studies of the University of Cambridge,* 5th ed. (London: Parker, 1850), 274 [emphasis added].

14. Louis Agassiz to Adam Sedgwick, Jun. 1845, in Agassiz, ed., *Louis Agassiz,* 392.

15. Louis Agassiz, "On the Succession and Development of Organized Beings at the Surface of the Terrestrial Globe; Being a Discourse Delivered at the Inauguration of the Academy of Neuchatel," *Edinburgh New Philosophical Journal,* 33 (1942), 399 [emphasis added].

16. Richard Owen, "Report on British Fossil Reptiles," in British Association for the Advancement of Science, *Report of the Eleventh Meeting* (London: John Murray, 1842), 204.

17. *Ibid.,* 202, 204.

18. Thomas Henry Huxley to W. MacLeay, 9 Nov. 1851, in Leonard Huxley, ed., *Life and Letters of Thomas Henry Huxley,* 1 (New York: Appleton, 1901), 101.

19. Charles Lyell, "Journal to Mary Horner," in Katherine Lyell, ed., *Life, Letters and Journals of Sir Charles Lyell, Bart.,* 1 (London: John Murray, 1881), 375.

20. Charles Lyell, "Extract 1," in Martin J. S. Rudwick, "Charles Lyell Speaks in the Lecture Theatre," *British Journal for the History of Science,* 9 (1976), 148.

21. Charles Lyell to Charles Darwin, 15 Mar. 1863, in Frederick

Burkhardt et al., eds., *The Correspondence of Charles Darwin*, 11 (Cambridge: Cambridge University Press, 1999), 230.

22. T. H. Huxley, "The Progress of Science," in T. H. Huxley, ed., *Method and Results: Essays* (New York: Appleton, 1898), 99.

23. Adam Sedgwick to David Livingstone, 16 Mar. 1865, in John Willis Clark and Thomas McKenny Hughes, eds., *The Life and Letters of the Reverend Adam Sedgwick*, 2 (Cambridge: Cambridge University Press, 1890), 412.

CHAPTER 3. ON THE ORIGINS OF DARWINISM

1. Charles Darwin, *Charles Darwin's Beagle Diary*, Richard Darwin Keynes, ed. (Cambridge: Cambridge University Press, 1988), 17–18.

2. Charles Darwin to J. S. Henslow, 5 Sept. 1831, in Frederick Burkhardt et al., eds., *The Correspondence of Charles Darwin*, 1 (Cambridge: Cambridge University Press, 1985), 142 [spelling corrected].

3. Charles Darwin to Charles Whitley, 9 Sept. 1831, in Burkhardt et al., eds., *The Correspondence of Charles Darwin*, vol. 1, 150.

4. Charles Darwin, *Journal of Researches into the Natural History and Geology of the Countries Visited During the Voyage of H. M. S. Beagle Round the World*, 2nd ed. (New York: Appleton, 1845), 1.

5. *Ibid.*, 5–6 [quote on 5].

6. Darwin, *Beagle Diary*, 23.

7. Charles Darwin, *The Autobiography of Charles Darwin, 1809–1882*, Nora Barlow, ed. (London: Collins, 1958), 77, 81.

8. Darwin, *Beagle Diary*, 292.

9. Darwin, *Journal of Researches*, 304–5 [emphasis in orginal].

10. Charles Darwin to W. D. Fox, Mar. 1835, in Burkhardt et al., eds., *The Correspondence of Charles Darwin*, vol. 1, 433.

11. Charles Lyell, *Principles of Geology*, 2 (London: John Murray, 1831), 124–25 [quote on 126; emphasis in original].

12. E.g., Darwin, *Beagle Diary*, 356.

13. *Ibid.*, v.

14. Charles Lyell, *Principles of Geology*, 4th ed., 1 (London: John Murray, 1835), 43.

15. Charles Darwin to Leonard Horner, 29 Sep. 1844, in Frederick

Burkhardt et al., eds., *The Correspondence of Charles Darwin,* 3 (Cambridge: Cambridge University Press, 1988), 55.

16. Charles Darwin, *Foundations of the Origin of Species: Two Essays Written in 1842 and 1844,* Francis Darwin, ed. (Cambridge: Cambridge University Press, 1909), 182. For a sample of Darwin's musing over the preceding eight years on this point relating to Galápagos species, see Charles Darwin, *Charles Darwin's Notebooks, 1836–1844,* Paul H. Barrett et al., eds. (Ithaca: Cornell University Press, 1987), 195, 296, 305, 405, 425, 640.

17. Charles Darwin, "Darwin's Journal," Gavin de Beer, ed., *Bulletin of British Museum (Natural History) Historical Series,* 2, no. 1 (1959), 7.

18. Charles Darwin, *The Red Notebook of Charles Darwin,* Sandra Herbert, ed. (Ithaca: Cornell University Press, 1980), 63 [spelling corrected].

19. Darwin, *Foundations,* 33.

20. Darwin, *Darwin's Notebooks,* 296.

21. Charles Darwin, "Charles Darwin and the Galápagos Islands," Nora Barlow, ed., *Nature,* 136 (1935), 391 [emphasis in original; spelling corrected].

22. Darwin, *Journal of Researches,* 373.

23. *Ibid.,* 225.

24. Darwin, *Darwin's Notebooks,* 264, 291, 542, 549–50, 558–59, 567, 574 [one quote from each page or pair of pages; emphasis in original; spelling corrected].

25. Thomas Robert Malthus, *An Essay on the Principle of Population,* 1st ed. (New York: Norton, 1976 rpt.), 20.

26. Darwin, *Darwin's Notebooks,* 375–76 [spelling corrected].

27. Darwin, *Autobiography,* 120.

28. Darwin, *Darwin's Notebooks,* 416.

29. Peter J. Bowler, *Evolution: The History of an Idea* (Berkeley: University of California Press, 1984), 158; Adrian Desmond and James Moore, *Darwin* (New York: Warner, 1991), 384; Randal Keynes, *Annie's Box: Charles Darwin, His Daughter and Human Evolution* (London: Fourth Estate, 2001), 245–53, 282.

30. Darwin, *Darwin's Notebooks,* 414 [spelling corrected].

31. Janet Browne, *Charles Darwin: Voyaging,* 1 (Princeton: Princeton University Press, 1995), 390.

32. Charles Darwin to J. D. Hooker, 13 Jul. 1856, in Frederick Burkhardt et al., eds., *The Correspondence of Charles Darwin*, 6 (Cambridge: Cambridge University Press, 1990), 178.

33. Duncan M. Porter, "On the Road to the *Origins* with Darwin, Hooker, and Gray," *Journal of the History of Biology*, 26 (1993), 3–8, 35.

34. Alfred Russel Wallace, "On the Tendency of Varieties to Depart Indefinitely from the Original Type," in Jane R. Camerini, ed., *The Alfred Russel Wallace Reader: A Selection of Writings from the Field* (Baltimore: Johns Hopkins University Press, 2002), 143, 148, 150 [emphasis in original].

35. Charles Darwin to Charles Lyell, 18 Jun. 1858, in Frederick Burkhardt et al., eds., *The Correspondence of Charles Darwin*, 7 (Cambridge: Cambridge University Press, 1991), 55 [spelling corrected].

CHAPTER 4. ENTHRONING NATURALISM

1. "Darwin on the Origin of Species," *The Times* [London], 26 Dec. 1859, 8. *Times* reviewer Samuel Lucas wrote the first two paragraphs of this review; T. H. Huxley wrote the rest. Here, quotes in first three sentences come from Lucas's portion of the review; others come from Huxley's portion.

2. *Ibid.*, 8–9.

3. John Fiske to Abby Brooks Fiske, 13 Nov. 1873, *The Personal Letters of John Fiske* (Cedar Rapids: Torch, 1939), 121–22.

4. T. H. Huxley to Charles Darwin, 23 Nov. 1859, in Huxley, ed., *Life and Letters of Thomas Henry Huxley*, vol. 1, 189. Huxley's use of the term "working hypothesis" for Darwin's theory is from Janet Browne, *Charles Darwin: The Power of Place*, 2 (New York: Knopf, 2002), 93.

5. Charles Darwin to T. H. Huxley, 25 Nov. 1859, in Burkhardt et al., eds., *The Correspondence of Charles Darwin*, vol. 7, 398.

6. The popular depiction of these four scientists as Darwinism's Four Musketeers is from Browne, *Charles Darwin*, vol. 2, 130.

7. T. H. Huxley, "Darwin on the Origin of Species," *Westminister Review*, 73 (1860), 295, 304–5 [all quotes from these pages].

8. Huxley to Charles Darwin, 188.

9. Huxley, "Darwin on Origin," 295. The depiction of *Origin of Species* as one long argument is captured in the title of a book by a leading twentieth-century Darwinian biologist: Ernst Mayr, *One Long Argument: Charles Darwin and the Genesis of Modern Evolutionary Thought* (Cambridge: Harvard University Press, 1991).

10. Charles Darwin, *On the Origin of Species,* 1st ed. (Cambridge: Harvard University Press, 1964 fasc. rpt.), 30.

11. *Ibid.,* 62, 63, 75, 80–81 [all quotes from these pages].

12. *Ibid.,* 170.

13. *Ibid.,* 13, 95–96, 194–95 [all quotes from these pages].

14. *Ibid.,* 111, 116 [quotes from both pages].

15. *Ibid.,* 126, 130 [quotes from both pages].

16. *Ibid.,* 456.

17. *Ibid.,* 489.

18. *Ibid.,* 490.

19. T. H. Huxley to Charles Kingsley, 22 May 1863, in Huxley, ed., *Life and Letters of Thomas Henry Huxley,* vol. 1, 263.

20. William Buckland, *Geology and Mineralogy Considered with Reference to Natural Theology,* 1 (London: Pickering, 1837) 237–40.

21. William Paley, *Natural Theology: or, Evidences of the Existence and Attributes of the Deity, Collected from the Appearances of Nature* (Philadelphia: John F. Watson, 1814), 16.

22. Darwin, *Origin of Species,* 200–201.

23. Adam Sedgwick to Charles Darwin, 24 Nov. 1959, in Burkhardt et al., eds., *The Correspondence of Charles Darwin,* vol. 7, 396–97 [emphasis in original].

24. Charles Darwin to Asa Gray, 22 May 1860, in Frederick Burkhardt et al., eds., *The Correspondence of Charles Darwin,* 8 (Cambridge: Cambridge University Press, 1993), 224.

25. Sedgwick to Darwin, 397.

26. T. H. Huxley, *Man's Place in Nature* (New York: Modern Library, 2001 rpt.), 106.

27. Browne, *Charles Darwin,* vol. 2, 221.

28. T. H. Huxley to Francis Darwin, 27 Jun. 1891, in Huxley, ed., *Life and Letters of Thomas Henry Huxley,* vol. 1, 202.

29. This and other versions of Huxley's response are collected in J. Vernon Jenson, "Return to the Wilberforce-Huxley Debate,"

British Journal for the History of Science, 21 (1988), 168. Although Huxley's ad lib became legend, it was not widely reported by the press covering the event.

30. Huxley, *Man's Place in Nature*, 112–13.
31. Charles Darwin, *The Descent of Man and Selection in Relation to Sex*, 1 (New York: Appleton, 1871), 3.
32. Adrian Desmond and James Moore, *Darwin* (New York: Warner Books, 1991), 203–4 [emphasis in original].
33. Darwin, *The Descent of Man*, vol. 1, 31, 203–4.
34. *Ibid.*, 55, 60, 65.
35. *Ibid.*, 140, 172–73.
36. Darwin, *The Descent of Man and Selection in Relation to Sex*, 2 (New York: Appleton, 1871), 311, 352, 355, 365.
37. Desmond and Moore, *Darwin*, 579 [emphasis in original].
38. Darwin, *The Descent of Man*, vol. 2, 387.
39. Alfred Russel Wallace, "Sir Charles Lyell on Geological Climates and the Origin of Species," *Quarterly Review* [American Edition], 126 (1869), 204, 205.

CHAPTER 5. ASCENT OF EVOLUTIONISM

1. Emma Darwin to Leonard Darwin, Sep. 1876, in Henrietta Litchfield, ed., *Emma Darwin: A Century of Family Letters, 1792–1896*, 2 (London: John Murray, 1915), 223.
2. Sketch reproduced in the gallery of illustrations between pages 460 and 461 of Desmond and Moore, *Darwin*.
3. John Fiske to Abby Brooks Fiske, 123.
4. Browne, *Charles Darwin*, vol. 2, 388–91.
5. Charles Darwin to Asa Gray, 11 May 1863, in Burkhardt et al., eds., *The Correspondence of Charles Darwin*, vol. 11, 403 [Darwin underlined "Creation" and "Modification" once and "or" twice].
6. Edward D. Cope, "Evolution and Its Consequences," *Penn Monthly*, 3 (1972), 223.
7. Peter J. Bowler, *Evolution*, 184.
8. Asa Gray, *Natural Science and Religion: Two Lectures Delivered to the Theological School of Yale College* (New York: Scribner, 1880), 61–62.

9. George F. Wright, "Recent Works Bearing on the Relation of Science to Religion: No. II," *Bibliotheca Sacra,* 33 (1876), 480.

10. Louis Agassiz, *Essay on Classification,* Edward Lurie, ed. (Cambridge: Harvard University Press, 1962), 44, 136–37.

11. Darwinian philosopher of science Michael Ruse expressed this view in Michael Ruse, *The Evolution Wars: A Guide to the Debates* (Santa Barbara: ABC-CLIO, 2000), 87.

12. Alfred Russel Wallace, *Island Life* (New York: Prometheus Books, 1998 fasc. rpt.), 4 [summarizing findings from earlier works].

13. *Ibid.,* 13.

14. Alfred Russel Wallace, "The 'Why' and the 'How' of Land Nationalization," *Macmillan's Magazine,* 48 (1883), 492.

15. Alfred Russel Wallace, "Evolution and Character," *Fortnightly Review,* 83 ns (1908), 23.

16. *Ibid.,* [emphasis added].

17. Alfred Russel Wallace, *Studies Scientific and Social,* 2 (London: Macmillan, 1900), 507.

18. Charles Darwin, in Francis Darwin, ed., *The Life and Letters of Charles Darwin,* 2 (New York: Appleton, 1897), 365.

19. Wallace, "Evolution and Character," 20.

20. Ernst Haeckel, *The History of Creation; or, the Development of the Earth and Its Inhabitants by the Action of Natural Causes,* 1st ed., 1 (New York: Appleton, 1876), 310–11.

21. Ernst Haeckel, *The Evolution of Man,* 2 (New York: Appleton, 1897), pl. xv [facing p. 188].

22. Fleeming Jenkin, "The *Origin of Species,*" *North British Review,* 46 (1867), 290.

23. Charles Darwin, *The Variation of Animals and Plants Under Domestication,* 2, 2nd ed. (London: John Murray, 1875), 373, 397 [quote].

24. Charles Darwin to Charles Lyell, 10 Dec. 1859, in Frederick Burkhardt et al., eds, *The Correspondence of Charles Darwin,* 4 (Cambridge: Cambridge University Press, 1991), 423.

25. John F. W. Herschel, *Physical Geography* (Edinburgh: Black, 1861), 12.

26. Vernon L. Kellogg, *Darwinism To-Day* (New York: Holt, 1907), 3, 5, 6 [includes Dennert quote].

CHAPTER 6. MISSING LINKS

1. "A Logical Refutation of Mr. Darwin's Theory," *Punch,* 1 Apr. 1871, 130.

2. Alfred Russel Wallace, "Evolution and Character," 22.

3. "That Troublesome Monkey Again," *Fun,* 16 Nov. 1872, 203.

4. Inscription and Darwin's response are reprinted in Browne, *Charles Darwin,* vol. 2, 403. Darwin never cut the pages of Marx's book.

5. Elizabeth Cady Stanton, *The Woman's Bible,* 2 (Boston: Northeastern University Press, 1993 rpt.), 214.

6. Andrew Carnegie, *Autobiography* (Boston: Houghton Mifflin, 1920), 339.

7. The poster titled "Our National Church" is reprinted as the end pages of Warren Sylvester Smith, *The London Heretics: 1870–1914* (New York: Dodd, Mead & Co., 1968).

8. T. H. Huxley, "On the Animals Which Are Most Nearly Intermediate Between Birds and Reptiles," *Annals and Magazine of Natural History,* ser. 4, vol. 2 (1868), 70, 73.

9. T. H. Huxley, *American Addresses, with a Lecture on the Study of Biology* (London: Macmillan, 1886), 90.

10. Charles Darwin to O. C. Marsh, 31 Aug. 1880, in Darwin, ed., *The Life and Letters of Charles Darwin,* vol. 2, 417.

11. O. C. Marsh, "Introduction and Succession of Vertebrate Life in America," *Nature,* 16 (1877), 471.

12. T. H. Huxley, *Man's Place in Nature,* 72.

13. *Ibid.,* 159, 166.

14. Charles Lyell, *The Geological Evidence of the Antiquity of Man* (London: Dent, 1914 rpt.), 393.

15. Charles Darwin to T. H. Huxley, 26 Feb. 1863, in Burkhardt et al., eds., *The Correspondence of Charles Darwin,* vol. 11, 181.

16. Charles Lyell to Charles Darwin, 11 Mar. 1863, Burkhardt et al., eds., *The Correspondence of Charles Darwin,* vol. 11, 218.

17. Ernst Haeckel, *The History of Creation; or the Development of the Earth and Its Inhabitants by the Action of Natural Causes,* 5th ed., 1, (New York: Appleton, 1911), 6 [emphasis in original].

18. For example, *ibid.,* 405–07.

19. Charles Darwin, *The Descent of Man,* vol. 1, 191.

20. Lyell, *The Geological Evidence*, 388.
21. Haeckel, *The History of Creation or the Development of the Earth and Its Inhabitants by the Action of Natural Causes*, 5th ed., 2, (New York: Appleton, 1911), 398–99, 436–37, 445 [quote].
22. From Eugène Dubois's third quarterly report of 1892, quoted at length in Pat Shipman, *The Man Who Found the Missing Link: Eugène Dubois's Thirty-Year Struggle to Prove Darwin Right* (New York: Simon & Schuster, 2001), 166.
23. Raymond A. Dart, *Adventures with the Missing Link* (New York: Harper, 1959), 5.
24. Raymond A. Dart, "Australopithecus africanus: The Man-Ape of South Africa," *Nature*, 115 (1925), 198.
25. Robert Broom, "Some Notes on the Taung Skull," *Nature*, 115 (1925), 571.
26. Dart, "Australopithecus," 198–99.

CHAPTER 7. GENETICS ENTERS THE PICTURE
1. Francis Galton to Charles Darwin, 24 Dec. 1869, in Karl Pearson, ed., *The Life, Letters and Labours of Francis Galton*, 1 (Cambridge: Cambridge University Press, 1914), pl. 2.
2. E.g., Francis Galton, "Hereditary Talent and Character," *Macmillan's Magazine*, 14 (1869), 325–26; Francis Galton, *Hereditary Genius: An Inquiry into Its Laws and Consequences* (New York: Appleton, 1887), 330–39.
3. For historical analysis of the socially constructed prejudices that Galton brought to his science, see Ruth Schwartz Cowan, *Francis Galton and the Study of Heredity in the Nineteenth Century* (New York: Garland, 1985), 64–69, 255–61; Raymond E. Fancher, "Francis Galton's African Ethnography and Its Role in the Development of His Psychology," *British Journal for the History of Science*, 16 (1983), 67–68, 79.
4. Galton, "Hereditary Talent," 322 [emphasis in original].
5. Hugo de Vries, translated in Erik Zevenhuizen, "The Hereditary Statistics of Hugo de Vries," *Acta botanica Neerlandica*, 47 (1998), 453.
6. A survey conducted by a Dutch newspaper in 1916 found that its readers ranked de Vries as the fourth most important Dutch cit-

izen of the preceding half century. Bert Theunissen, "The Scientific and Social Context of Hugo de Vries' *Mutationstheorie*," *Acta botanica Neerlandica*, 47 (1998), 487.

7. Hugo de Vries to Jan Willem Moll, 2 Apr. 1903, translated in Ida H. Stamhuis et al., "Hugo de Vries on Heredity, 1889–1903," *Isis*, 90 (1999), 255.

8. Gregor Mendel, quoted in Maitland A. Edey and Donald C. Johanson, *Blueprints: Solving the Mystery of Evolution* (New York: Penguin, 1990), 121.

9. Gregor Mendel, "Experiments on Plant Hybrids," rpt. in Curt Stern and Eva R. Sherwood, eds., *The Origins of Genetics: A Mendel Source Book* (San Francisco: Freeman, 1966), 5–7 [quotes on p. 5].

10. Thomas Hunt Morgan, "For Darwin," *Popular Science Monthly*, 74 (1909), 380.

11. Thomas Hunt Morgan, *Evolution and Adaptation* (New York: Macmillan, 1903), 165–66, 260.

12. Thomas Hunt Morgan to Hans Driesch, 23 Nov. 1910, in Morgan-Driesch correspondence, microfilm copy in the library of the American Philosophical Society, Philadelphia.

CHAPTER 8. APPLIED HUMAN EVOLUTION

1. Francis Galton, *Inquiries into Human Faculty and Its Development* (London: Macmillan, 1883), 1.

2. Francis Galton, *Essays in Eugenics* (London: Eugenics Education Society, 1909), 24–25.

3. This sentence is compiled from two parallel passages in Galton's voluminous writings: Francis Galton, "Hereditary Talent and Character," 163 [first quote]; Galton, *Inquiries into Human Faculty*, 217 [second quote].

4. Galton, "Hereditary Talent and Character," 165.

5. Francis Galton, *Memories of My Life* (London: Methuen, 1908), 315–16.

6. Galton, "Hereditary Talent and Character," 319 [quote]; Francis Galton, "Hereditary Improvement," *Fraser's Magazine*, 87 (1973), 125–28 [suggests sexual segregation].

7. Galton, "Hereditary Talent and Character," 319–20.

8. *Ibid.*, 165–66.

9. Galton, *Memories of My Life,* 323.

10. Galton, *Inquiries into Human Faculty,*" 200.

11. Charles Darwin, *The Descent of Man,* vol. 1, 228–31 [in subchapter entitled, "On the Extinction of the Races of Man"].

12. R. L. Dugdale, *The Jukes: A Story in Crime, Pauperism, Disease, and Heredity,* 5th ed. (New York: Putnam, 1895), 7–15, 69–70.

13. Darwin, *The Descent of Man,* vol. 1, 106–7.

14. Dugdale, *The Jukes,* 55, 57, 65.

15. This particular phrase appeared in an unsigned editorial published in Margaret Sanger's *Birth Control Review,* but it was broadly representative of early-twentieth-century eugenic thought. "Intelligent or Unintelligent Birth Control?" *Birth Control Review,* May 1919, 12.

16. Thomas Robert Malthus, *An Essay on the Principle of Population,* 1st ed. (New York: Norton, 1976 rpt.), 29 ["struggle for existence"], 58 ["goad of necessity"], 64. Mathus did not envision these processes as permanently improving the species, however, because he regarded basic human nature as fixed since Creation.

17. Cesare Lombroso, *Crime: Its Causes and Remedies* (Boston: Little, Brown, 1911), 365–75.

18. Herbert Spencer, *The Man Versus the State, with Four Essays on Politics and Society* (Baltimore: Penguin, 1969 rpt.), 82.

19. William Graham Sumner, "Sociology," in Stow Persons, ed., *Social Darwinism: Selected Essays of William Graham Sumner* (Englewood Cliffs, NJ: Prentice-Hall, 1963), 16–17.

20. *Lochner v. New York,* 198 U.S. 45, 75 (1905) (Holmes, J., dissenting).

21. Herbert Spencer, *Education: Intellectual, Moral and Physical* (New York: Appleton, 1860), 213.

22. Ernst Haeckel, *The History of Creation,* 5th ed., vol. 2, 445.

23. Joseph Le Conte, *The Race Problem in the South* (New York: Appleton, 1892), 367. For a similar comment by Sumner, see William Graham Sumner, *Collected Essays in Political and Social Science* (New York: Holt, 1885), 130.

24. Charles Darwin, *The Descent of Man,* vol. 1, 229.

25. Georges Vacher de Lapouge, *L'Aryen: son rôle social* (Paris: Fontemoing, 1899), 512.

26. Haeckel, *The History of Creation,* 5th ed., vol. 1, 20, 175 [emphasis in original].

27. Vernon Kellogg, *Headquarters Nights: A Record of Conversations and Experiences at the Headquarters of the German Army in France and Belgium* (Boston: Atlantic Monthly Press, 1917), 22–29.

28. See note 15 above.

29. Arthur H. Estabrook, *The Jukes in 1915* (Washington: Carnegie Institution, 1916), 85.

30. Henry Herbert Goddard, *The Kallikak Family: A Study in the Heredity of Feeble-Mindedness* (New York: Macmillan, 1913), 60.

31. H. H. Goddard, "Four Hundred Feeble-Minded Children Classified by the Binet Method," *Journal of Psycho-Asthenics,* 15 (1910), 17, 26–27.

32. H. L. Mencken, "Utopia by Sterilization," *American Mercury,* 41 (1937), 399, 405.

33. Abraham Myerson et al., *Eugenical Sterilization: A Reorientation of the Problem* (New York: Macmillan, 1936), 179.

34. Georges Vacher de Lapouge, "L'anthropologie et la science politique," *Revue d'anthropologie,* 16 (1887), 140.

35. Lists of state sterilization laws, giving either their dates of enactment or the number of sterilizations performed, are in Moya Woodside, *Sterilization in North Carolina: A Sociological and Psychological Study* (Chapel Hill: University of North Carolina Press, 1950), 194–95; Jonas Robitscher, ed., *Eugenic Sterilization* (Springfield, IL: Thomas, 1973), 118–19.

36. Erwin Baur, quoted in Max Weinreich, *Hitler's Professors: The Part of Scholarship in Germany's Crimes Against the Jewish People* (New Haven: Yale University Press, 1999 rpt.), 31 [emphasis in original]. Baur's support for eugenic sterilization was not an isolated example among German biologists. Academic biologists joined the Nazi party in Germany at a higher rate than any other professional group; more than half of them became members. "That so many joined the Party (and also joined the SS and SA) is not explained by pressure," political scientist Diane Paul concludes. "It rather reflects their enthusiasm for a regime that finally gave biologists, and geneticists in particular, the support that they thought was their due. Far from being repressed, genetics—

which was considered to be of great ideological, military, and economic importance to the regime—flourished in the Third Reich." Diane B. Paul, *Controlling Human Heredity: 1865 to the Present* (Atlantic Highlands, NJ: Humanities Press, 1995), 91.

37. Eugenics Record Office, *Bulletin 10B: Legal, Legislative and Administrative Aspects of Sterilization* (Cold Spring Harbor, NY: Eugenics Record Office, 1914) 144–46.

38. *Buck v. Bell*, 274 U.S. 200, 205 (1927).

39. George William Hunter, *A Civic Biology* (New York: American, 1914), 261–63. Regarding the popularity of this textbook, see Edward J. Larson, *Summer for the Gods: The Scopes Trial and America's Continuing Debate Over Science and Religion* (New York: Basic Books, 1997), 23.

40. The movie was first released in 1916 under the title *The Black Stork* but survives only in its 1927 form, *Are You Fit to Marry?* All quotes are from a VHS tape of the movie *Are You Fit to Marry?* (Quality Amusement Corp., 1927), John E. Allen Archives, Nebraska ETV Network, Lincoln, NE. Since the movie is silent, with all words printed as subtitles, the quotes can be read from the screen. The origins, contents, and reception of the movie are discussed in Martin S. Pernick, *The Black Stork: Eugenics and the Death of "Defective" Babies in American Medicine and Motion Pictures since 1915* (New York: Oxford University Press, 1996), 143–58. Pernick's book also contains many quotes from the movie.

41. G. K. Chesterton, *Eugenics and Other Evils* (London: Cassell, 1922), 180.

42. As late as 1930, the influential American geneticist Edwin G. Conklin still asserted that "all modern geneticists approve the segregation or sterilization of those who are known to have serious hereditary defects, such as hereditary feeblemindedness, insanity, etc." Edwin G. Conklin, "The Purposive Improvement of the Human Race," in E. V. Cowdry, ed., *Human Biology and Population Improvement* (New York: Hoeber, 1930), 577.

CHAPTER 9. AMERICA'S ANTI-EVOLUTION CRUSADE

1. "'Where There Is No Vision the People Perish'—Text of Sunday's Sermon," *Commercial Appeal* [Memphis], 6 Feb. 1925, 9. For a

general description of Sunday's style in delivering this and other sermons, see Ridley Wills, "Huge Throng Joins in Welcoming Sunday at Opening Meeting," *Commercial Appeal* [Memphis], 6 Feb. 1925, 1; and William G. McLaughlin, Jr., *Billy Sunday Was His Real Name* (Chicago: University of Chicago Press, 1955), 154–88.

2. McLaughlin, *Billy Sunday*, 151.

3. William T. Ellis, *"Billy" Sunday: The Man and his Message* (Philadelphia: Universal Books, 1914), 139.

4. From Sunday's sermons, quoted in McLaughlin, *Billy Sunday*, 121, 132.

5. Ridley Wills, "Church School Today Hotbed of Infidelity, Billy Sunday Charges," *Commercial Appeal* [Memphis], 18 Feb. 1925, 2.

6. Charles Hodge, *What Is Darwinism?* (New York: Scribner, 1974), 11, 173.

7. Dwight L. Moody, *Moody's Latest Sermons* (New York: Revell, 1900), 59.

8. George Herbert Betts, *The Beliefs of 700 Ministers and Their Meaning for Religious Education* (New York: Abingdon Press, 1929), 26, 44.

9. Henry Ward Beecher, *Evolution and Religion* (Boston: Pilgrim Press, 1885), 50–53 [emphasis in original].

10. William Jennings Bryan, "In the Chicago Convention," in William Jennings Bryan, ed., *Speeches of William Jennings Bryan*, 1 (New York: Funk & Wagnalls, 1909), 249.

11. William Jennings Bryan, "The Prince of Peace," in William Jennings Bryan, ed., *Speeches of William Jennings Bryan*, 2 (New York: Funk & Wagnalls, 1909), 268.

12. William Jennings Bryan, *In His Image* (New York: Revell, 1922), 94, 98, 100, 125.

13. William Jennings Bryan, quoted in Lawrence W. Levine, *Defender of the Faith: William Jennings Bryan, The Last Decade, 1915–1925* (New York: Oxford University Press, 1965), 277.

14. William Jennings Bryan and Mary Baird Bryan, *The Memoirs of William Jennings Bryan* (Philadelphia: United, 1925), 179–80.

15. William Jennings Bryan, "Speech to Legislature," in William Jennings Bryan, *Orthodox Christianity Versus Modernism* (New York: Revell, 1923), 46 [emphasis in original].

16. "Are People People?" *Chicago Tribune*, 20 Jun. 1923, 8.

17. 1923 Fla. House Concurrent Resolution 7.

18. Ridley Wills, "Huge Throng Joins in Welcome," 1.

19. Howard Eskridge, "Senate Passes Evolution Bill," *Nashville Banner*, 13 Mar. 1925, 10.

20. William Jennings Bryan to Austin Peay, 24 Mar. 1925, in William Jennings Bryan Papers, Library of Congress, Washington, D.C.

21. "Plan Assault on State Law on Evolution," *Chattanooga Daily Times*, 4 May 1925, 5 [ACLU press release as it appeared in the newspaper read by Dayton civic leaders].

22. H. L. Mencken, "The Monkey Trial: A Reporter's Account," in Jerry R. Tomkins, ed., *D-Days at Dayton: Reflections of the Scopes Trial* (Baton Rouge: Louisiana State University Press, 1965), 35 [reprint of 9 Jul. 1925 column].

23. Arthur Garfield Hays, "The Strategy of the Scopes Defense," *The Nation*, 5 Aug. 1925, 158.

24. William Jennings Bryan, in *The World's Most Famous Court Case: Tennessee Evolution Case* (Dayton, TN: Bryan College, 1990), 288 [trial transcript].

25. Clarence Darrow and William Jennings Bryan, in *The World's Most Famous Court Case*, 302.

26. Arthur Garfield Hays, *Let Freedom Ring* (New York: Liveright, 1928), 77.

27. William Jennings Bryan, in *The World's Most Famous Court Case*, 299.

28. "Ended at Last," *The New York Times*, 22 Jul. 1925, 18.

CHAPTER 10. THE MODERN SYNTHESIS

1. For a discussion of early interpretations of the peppered-moth phenomenon, see Michael E. N. Majerus, *Melanism: Evolution in Action* (New York: Oxford University Press, 1998), 99–104.

2. J.B.S. Haldane, "A Mathematical Theory of Natural and Artificial Selection: Part I," *Transactions of the Cambridge Philosophical Society*, 23 (1924), 19.

3. Ronald William Clark, *J.B.S.: The Life and Work of J.B.S. Haldane* (Oxford: Oxford University Press, 1984), 65.

4. Haldane, "A Mathematical Theory," 26.

5. J.B.S. Haldane, *The Causes of Evolution* (Princeton: Princeton University Press, 1990 rpt.), 1, 92.

6. *Ibid.*, 88.

7. R. A. Fisher, "Mendelism and Biometry," in J. H. Bennett, ed., *Natural Selection, Heredity, and Eugenics: Including Selected Correspondence of R. A. Fisher with Leonard Darwin and Others* (Oxford: Clarendon Press, 1983), 53–54.

8. All quotes are taken from Wright's classic 1931 paper and influential 1932 metaphorical elaboration of it. Sewall Wright, "Evolution in Mendelian Populations," *Genetics*, 16 (1931), 97–159; Sewall Wright, "The Roles of Mutation, Inbreeding, Crossbreeding and Selection in Evolution," *Proceedings of the Sixth International Congress of Genetics*, 1 (1932), 356–66. These and other significant papers by Wright are reprinted in Sewall Wright, *Evolution: Selected Papers*, William B. Provine, ed. (Chicago: University of Chicago Press, 1986).

9. "In a large population, divided and subdivided into partially isolated local races of small size, there is a continually shifting differentiation among the later," Wright explained in his classic 1931 paper, "which inevitably brings about...rapid evolution. Complete isolation in this case...originates new species differing for the most part in nonadaptive respects." Wright, "Evolution in Mendelian Populations," 158.

10. Sewall Wright, "The Genetical Theory of Natural Selection: A Review," *Journal of Heredity*, 21 (1930), 350.

11. R. A. Fisher, "A Review of 'Evolution in Mendelian Populations' (S. Wright, 1931)," *Eugenics Review*, 23 (1932), 90.

12. Theodosius Dobzhansky, "The Birth of the Genetic Theory of Evolution in the Soviet Union in the 1920s," in Ernst Mayr and William B. Provine, eds., *The Evolutionary Synthesis: Perspectives on the Unification of Biology* (Cambridge: Harvard University Press, 1980), 235.

13. Theodosius Dobzhansky, quote from personal letter in William B. Provine, "The Origins of the Genetics of Natural Populations Series," in R. C. Lewontin et al., eds., *Dobzhansky's Genetics of Natural Populations I–XLIII* (New York: Columbia University Press, 1981), 56.

14. Theodosius Dobzhansky, *Genetics and the Origin of Species* (New York: Columbia University Press, 1937), 133–34.

15. Stephen Jay Gould, "The Hardening of the Modern Synthesis," Marjorie Grene, ed., *Dimensions of Darwinism: Themes and Counterthemes in Twentieth-Century Evolutionary Theory* (Cambridge: Cambridge University Press, 1983), 73–80 [Dobzhansky], 83–86 [Wright].

16. Theodosius Dobzhansky, Oral History Memoir, 398–99, Columbia University, Oral History Research Office, New York, quoted in William B. Provine, *Sewall Wright and Evolutionary Biology* (Chicago: University of Chicago Press, 1986), 345.

17. Ernst Mayr, *Systematics and the Origin of Species* (New York: Columbia University Press, 1942), 120.

18. *Ibid.,* 155 [original in italics].

19. G. Ledyard Stebbins, Jr., *Variation and Evolution in Plants* (New York: Columbia University Press, 1950), 152.

20. Theodosius Dobzhansky, "Nothing in Biology Makes Sense Except in Light of Evolution," *American Biology Teacher,* 35 (1973), 125.

21. Harry S. Swarth, "The Avifauna of the Galápagos Islands," *Occasional Papers of the California Academy of Sciences,* 18 (1931), 19.

22. P. R. Lowe, "The Finches of the Galápagos in Relation to Darwin's Conception of Species," *Ibis,* 78 (1936), 310.

23. *Ibid.,* 320–21 [emphasis in original].

24. David Lack, *Darwin's Finches* (Cambridge: Cambridge University Press, 1983 rpt.), 1.

25. David Lack, "Evolution of the Galapagos Finches," *Nature,* 146 (1940), 326.

26. G. F. Gause, "Discussion of Paper by Thomas Park," *American Midland Naturalist,* 21 (1939), 255.

27. *Ibid.,* 62.

28. Lack, *Darwin's Finches,* 113, 159, 162. As noted above, Lowe originally proposed the name "Darwin's finches" for these birds in 1935, but the name did not stick until Lack used it in his 1947 book. Even Lack called them "Galapagos finches" in prior publications.

29. B. Rosemary Grant and Peter R. Grant, "Evolution of Darwin's Finches Caused by a Rare Climatic Event," *Proceedings of the Royal Society of London B,* 251 (1993), 111.

30. Peter R. Grant, "Natural Selection and Darwin's Finches," *Scientific American*, 273 (Oct. 1991), 86.

31. The mystique surrounding the peppered moth increased during the 1950s when British biologist Bernard Kettlewell attempted to show by field experiments that predation could cause differential selection at the rate calculated by Haldane as sufficient to transform the local population from virtually all speckled to virtually all black in fifty years. In his experiments, Kettlewell placed speckled and black moths in exposed position on blackened tree trunks and then noted the rate at which birds fed on them. Although his findings supported the selectionist hypothesis and were widely reported, they were suspect because peppered moths do not naturally rest on exposed tree trunks. For a critical analysis of Kettlewell's experiments, see Majerus, *Melanism*, 116–25.

CHAPTER 11. MODERN CULTURE WARS

1. For an example of Huxley's advocacy of what he called "evolutionary humanism" through his public positions, see Julian Huxley, *UNESCO: Its Purpose and Its Philosophy* (Washington: Public Affairs Press, 1947), 62.

2. Michael Ruse, *Mystery of Mysteries: Is Evolution a Social Construction?* (Cambridge: Harvard University Press, 1999), 94. Historian of science Daniel Kevles notes, "To Julian Huxley, however, it was incumbent upon mankind...to struggle to see that evolution continued to flourish. Embracing that aim, human beings could forge an evolutionary ethics, which would start with the principle that it was right to realize ever-new possibilities in evolution and which would consist of further principles extracted from what was necessary for the evolutionary process to proceed." Daniel J. Kevles, "Huxley and the Popularization of Science," in C. Kenneth Waters and Albert Van Helden, eds., *Julian Huxley: Biologist and Statesman of Science* (Houston: Rice University Press, 1992), 249. With respect to these principles, Kevles adds, "Huxley here appeared to be deluding himself. The seemingly cosmic objectivity of his system was to a considerable ex-

tent shaped by ... his resistance to that scourge of the twentieth century, totalitarianism." *Ibid.,* 250.

3. Huxley's earliest complete exposition of these views appeared in his aptly titled book, *Religion Without Revelation* (New York: Harper, 1927).

4. Julian Huxley, "The Humanist Frame," in Julian Huxley, ed., *The Humanist Frame* (New York: Harper, 1961), 17–18, 26 [reprint of 1959 Chicago address].

5. George Gaylord Simpson, *This View of Life: The World of an Evolutionist* (New York: Harcourt, Brace & World, 1964), 10, 25, 36 [partly drawn from earlier publications tied to the Darwinism centennial].

6. For example, Huxley, "The Humanist Frame," 17, 26; George Gaylord Simpson, *Concession to the Improbable: An Unconventional Autobiography* (New Haven: Yale University Press, 1978), 30.

7. Theodosius Dobzhansky to John Greene, 23 Nov. 1961 rpt. in Michael Ruse, *The Evolution Wars: A Guide to the Debates* (Santa Barbara: ABC-CLIO, 2000), 357.

8. David Lack, *Evolutionary Theory and Christian Belief: The Unresolved Conflict* (London: Methuen, 1961), 114.

9. George Gallup, Jr., and D. Michael Lindsay, *Surveying the Religious Landscape: Trends in U.S. Belief* (Harrisburg, PA: Morehouse, 1999), 38. National polling on this issue began in the 1980s, but trends suggest that similar results would have been found earlier.

10. Simpson, *View of Life,* 35–36 [based on his 1961 article in *Teachers College Record*].

11. For statistics on denominational growth rates, including analysis tying these statistics to theological liberalism, see Dean R. Hoge, "A Test of Theories of Denominational Growth and Decline," in Dean R. Hoge and David A. Roosen, eds., *Understanding Church Growth and Decline: 1950–1978* (New York: Pilgrim Press, 1979), 185–87. For raw data on denominational membership in 1925 compared to 1960, see "United States Religious Statistics," *World Almanac and Book of Facts for 1925* (New York: New York World, 1925), 360; "Census of Religious Bodies in United States," *World Almanac and Book of Facts for 1961* (New York: New York

World–Telegram, 1961), 695–96. For data suggesting support for creationism among members of Southern Baptist, Missouri Synod Lutheran, Mormon, Assembly of God, and Seventh-day Adventist churches, see Ronald L. Numbers, *The Creationists* (New York: Knopf, 1992), 300, 308.

12. Survey data on shifting church membership figures is discussed in George Gallup, Jr., and Jim Castelli, *The People's Religion: American Faith in the 90s* (New York: Macmillan, 1989), 29–30.

13. Statistics on membership growth and geographic spread of the Southern Baptist Convention appear in Phillip Barron Jones, "An Examination of the Statistical Growth of the Southern Baptist Convention," in Hoge and Roosen, eds., *Understanding Church Growth*, 160–63.

14. House Bill No. 301 (Arizona, 1964); Senate Bill No. 178 (Arizona, 1965). Neither bill passed. For a discussion of the opposition in Arizona to the BSCS textbooks written by the director of the BSCS project, see Arnold B. Grobman, *The Changing Classroom: The Role of the Biological Sciences Curriculum Study* (Garden City, NY: Doubleday, 1969), 205–07.

15. Based on directly applicable Supreme Court precedent involving classroom Bible teaching, a federal appellate court struck down the Tennessee creationism law in *Daniel v. Waters*, 515 F.2d 485 (6th Cir. 1975). The Tennessee law mandated that biology textbooks presenting any theory of human origins give "an equal amount of emphasis" to other theories "including, but not limited to, the Genesis account in the Bible." 1973 Tenn. Pub. Acts 377.

16. Dobzhansky to Greene, 357.

17. William Jennings Bryan, in *The World's Most Famous Court Case*, 302 [trial transcript].

18. C. I. Scofield, *The Scofield Reference Bible* (New York: Oxford University Press, 1909), 3, n. 2. ("But three *creative* acts of God are recorded in this chapter: (1) the heavens and the earth, vs. 1; (2) animal life, vs. 21; and (3) human life, vss. 26, 27. The first creative act refers to the dateless past, and gives scope for all the geologic ages.")

19. Clarence Darrow, in *The World's Most Famous Court Case*, 297.

20. Henry M. Morris and John D. Morris, *The Modern Creation Trilogy*, 1 (Green Forest, AR: Master, 1996), 95.
21. John C. Whitcomb, Jr., and Henry M. Morris, *The Genesis Flood: The Biblical Record and Its Scientific Significance* (Philadelphia: Presbyterian and Reformed Publishing Co., 1961), 16–20, 173–74, 233–39, 327–28.
22. Institute for Creation Research (ICR), *Scientific Creationism* (San Diego: Creation-Life Publishers, 1974), 255.
23. Ronald L. Numbers, *Darwinism Comes to America* (Cambridge: Harvard University Press, 1998), 5–6.
24. Henry M. Morris, "Introduction," in Henry M. Morris and Donald H. Rohrer, eds., *Creation: The Cutting Edge* (San Diego: Creation-Life Publishers, 1982), 9 [emphasis in original].
25. Henry M. Morris and John D. Morris, *The Modern Creation Trilogy*, 3 (Green Forest, AR: Master, 1996), 57–179 [quote on 129].
26. ICR, "Resolution for Balanced Presentation of Evolution and Scientific Creationism," *ICR Impact Series*, no. 71 (May 1979), ii.
27. ICR, *Scientific Creationism*, 5–14, 37–42, 54–69, 78–90, 111–29 [quote on 9].
28. A 1981 nationwide public-opinion survey commissioned by NBC News and the Associated Press found that 76 percent of the respondents favored teaching both creationism and the theory of evolution, 10 percent favored teaching only creationism, and 8 percent favored teaching only the theory of evolution, with 6 percent undecided. "76% for Parallel Teaching of Creation Theories," *San Diego Union*, 18 Nov. 1981, A15. Similar results are reported in *Williamsburg Charter Survey on Religion in Public Life: List of Tables* (Washington: Williamsburg Charter Foundation, 1988), table 35.
29. Ronald Reagan, in Kenneth M. Pierce, "Putting Darwin Back in the Dock," *Time*, 16 Mar. 1981, 80.
30. From the title of both statutes: 1981 Ark. Acts 590; 1981 La. Acts 685.
31. Frank White, in Andrew Polin, "Law Requires Schools Offer Theory Choice," *Arkansas Democrat*, 20 Mar. 1981, 1.
32. Roger Baldwin, in "Remember Scopes Trial? ACLU Does," *Times-Picayune/States-Item* (New Orleans), 22 Jul. 1981, 1.

33. 1981 Ark. Acts 590, § 4 (a).

34. *McLean v. Arkansas Board of Education,* 529 F. Supp. 1255, 1264, 1266, 1272 (E.D. Ark. 1982) [emphasis in original].

35. *Aguillard v. Treen,* No. 81-4787 (E.D. La. 10 Jan. 1985), 9–10.

36. *Edwards v. Aguillard,* 482 U.S. 578, 592 (1987).

37. *Ibid.,* 602–3 (Powell, J., concurring).

38. For clear consensus assertion of this principle made on behalf of the prestigious National Academy of Sciences in a booklet published to counter the so-called "intelligent design" movement, see National Academy of Sciences, *Teaching About Evolution and the Nature of Science* (Washington: National Academy Press, 1998), 27–42.

39. Phillip E. Johnson, quoted in Tim Stafford, "The Making of a Revolution," *Christianity Today,* 8 Dec. 1997, 18 [emphasis in original].

40. Phillip E. Johnson, *Darwin on Trial* (Washington: Regnery, 1991), 45–72, 111–22; Phillip E. Johnson, *Defeating Darwinism by Opening Minds* (Dowers Grove, IL: InterVarsity Press, 1997), 53–67.

41. Michael J. Behe, *Darwin's Black Box: The Biochemical Challenge to Evolution* (New York: Simon & Schuster, 1996), 74–97.

42. For example, Kenneth R. Miller, *Finding Darwin's God: A Scientist's Search for Common Ground Between God and Evolution* (New York: HarperCollins, 1999), 152–58; Michael Ruse, *Can a Darwinian Be a Christian? The Relationship Between Science and Religion* (Cambridge: Cambridge University Press, 2001), 116–20.

43. Kenneth R. Miller, "'Intelligent Design' Nothing but Scientific Imposter," *Kansas City Star,* 1 Jul. 2000, B8. On the persuasiveness of Dembski's arguments, compare William A. Dembski, "Reinstating Design Within Science," *Rhetoric & Public Affairs,* 1 (1998), 515; with Celeste Michelle Condit, "The Rhetoric of Intelligent Design: Alternatives for Science and Religion," *Rhetoric & Public Affairs,* 1 (1998), 595.

44. Lawrence S. Learner, *State Science Standards: An Appraisal of Science Standards in 36 States* (Washington: Fordham Foundation, 1998), 14 [quoting from Alabama science standards].

45. *147 Congressional Record* S113377-78 (daily ed. 18 Dec. 2001) [statement of Senator Santorum on conference report]; Phillip

E. Johnson to Michael Ruse, 19 Dec. 2001 [e-mail message acknowledging Johnson's role in drafting Santorum's resolution and commenting on conference report—copy in author's file].

CHAPTER 12. POSTMODERN DEVELOPMENTS

1. James D. Watson, *The Double Helix: A Personal Account of the Discovery of the Structure of DNA* (New York: Norton, 1980 rpt.), 115.
2. J. D. Watson and F.H.C. Crick, "A Structure for Deoxyribose Nucleic Acid," *Nature,* 171 (1953), 738; J. D. Watson and F.H.C. Crick, "Genetical Implications of the Structure of Deoxyribonucleic Acid," *Nature,* 171 (1953), 967.
3. Edward O. Wilson, *Naturalist* (Washington: Island Press, 1994), 223–24.
4. *Ibid.,* 219.
5. Watson, *Double Helix,* 46.
6. R. C. Lewontin and J. L. Hubby, "A Molecular Approach to the Study of Genic Heterozygosity in Natural Populations. II. Amount of Variation and Degree of Heterozygosity in Natural Populations of *Drosophila pseudoobscura,*" *Genetics,* 54 (1966), 605–8; R. C. Lewontin, *The Genetic Basis of Evolutionary Change* (New York: Columbia University Press, 1974), 197–98, 267.
7. Wilson, *Naturalist,* 225.
8. W. D. Hamilton, *Narrow Roads of Gene Land: The Collected Papers of W. D. Hamilton,* 1 (Oxford: Freeman, 1996), 12.
9. W. D. Hamilton, "The Genetical Evolution of Social Behavior, I," *Journal of Theoretical Biology,* 7 (1964), 16.
10. W. D. Hamilton, "The Genetical Evolution of Social Behavior, II," *Journal of Theoretical Biology,* 7(1964), 31.
11. Hamilton, "The Genetical Evolution, I," 16.
12. Wilson, *Naturalist,* 319.
13. Edward O. Wilson, *Sociobiology: The New Synthesis* (Cambridge: Harvard University Press, 2000 rpt.), 547, 562.
14. Edward O. Wilson, *On Human Nature* (Cambridge: Harvard University Press, 1978), 167.
15. T. H. Huxley, "Evolution and Ethics," rpt. in Alan P. Barr, ed., *The Major Prose of Thomas Henry Huxley* (Athens, GA: University of Georgia Press, 1997), 327.

16. Theodosius Dobzhansky, "Anthropology and the Natural Sciences—the Problem of Human Evolution," *Current Anthropology,* 4 (1963), 146.

17. Elisabeth Allen et al., "Against 'Sociobiology,'" *The New York Review of Books,* 13 Nov. 1975, 43.

18. Wilson, *Sociobiology,* vii.

19. Richard Dawkins, "Foreword," in W. D. Hamilton, *Narrow Roads of Gene Land: The Collected Papers of W. D. Hamilton,* 2 (Oxford: Oxford University Press, 2001), xv.

20. Richard Dawkins, *The Selfish Gene* (New York: Oxford University Press, 1989 rpt.), v.

21. Richard Dawkins, *Climbing Mount Improbable* (New York: Norton, 1996) [quote from the title].

22. Richard Dawkins, *The Blind Watchmaker: Why the Evidence of Evolution Reveals a Universe Without Design* (New York: Norton, 1986), 4–5.

23. Wilson, *On Human Nature,* 192.

24. Michael Ruse and Edward O. Wilson, "The Evolution of Ethics," *New Scientist,* 17 Oct. 1985, 52.

25. Edward O. Wilson, *Consilience: The Unity of Knowledge* (New York: Knopf, 1998), 264–65. See also excerpts from Wilson interviews in Michael Ruse, *Monad to Man: The Concept of Progress in Evolutionary Biology* (Cambridge: Harvard University Press, 1996), 516.

26. Ruse, *The Evolution Wars,* 284.

27. Hamilton, *Narrow Roads,* vol. 1, 15.

28. Hamilton, *Narrow Roads,* vol. 2, liii.

29. Jon Seger and W. D. Hamilton, "Parasites and Sex," in Richard E. Michod and Bruce R. Levin, eds., *The Evolution of Sex: An Examination of Current Ideas* (Sunderland, MA: Sinauer, 1988), 176.

30. Hamilton, *Narrow Roads,* vol. 1, 17.

31. Hamilton, *Narrow Roads,* vol. 2, xlvii. Hamilton's reference to diabetics appeared later in this book, at page 456.

32. See, e.g., Stephen Jay Gould, *The Mismeasure of Man* (New York: Norton, 1981).

33. For Hamilton's comments on Gould and like-minded opponents of sociobiology, see Hamilton, *Narrow Roads,* vol. 2, 490–91. For Ruse's reference to Gould's Jewish roots, see Ruse, *The Evolution*

Wars, 245: "[Gould] saw [hereditarianism] as a threat to all that he held sacred and something to be opposed with all his might."

34. S. J. Gould and R. C. Lewontin, "The Spandrels of San Marco and the Panglossian Paradigm: A Critique of the Adaptationist Programme," *Proceedings of the Royal Society of London,* ser. B, vol. 205 (1979), 587.

35. Stephen Jay Gould, *Full House: The Spread of Excellence from Plato to Darwin* (New York: Harmony, 1996), 216.

36. Robert Wright, "The Accidental Creationist," *New Yorker,* 13 Dec. 1999, 64.

37. Stephen Jay Gould, "Opus 2000," *Natural History,* Aug. 1991, 14.

38. Niles Eldredge and Stephen Jay Gould, "Punctuated Equilibria: An Alternative to Phyletic Gradualism," in Thomas J. M. Schopf, ed., *Models in Paleobiology* (San Francisco: Freeman, Cooper, 1972), 84.

39. Niles Eldridge, *Reinventing Darwin: The Great Debate at the High Table of Evolutionary Thought* (New York: Wiley, 1995), 86.

40. *Ibid.,* 87.

41. For example, Stephen Jay Gould, *The Structure of Evolutionary Theory* (Cambridge: Harvard University Press, 2002), 774–784.

42. *Ibid.,* 99.

43. John Maynard Smith, "Genes, Memes, and Minds," *The New York Review of Books,* 30 Nov. 1995, 46.

44. Wright, "The Accidental Creationist," 56.

45. Francis S. Collins, "Human Genetics," in John F. Kilner et al., eds., *Cutting Edge of Bioethics* (Grand Rapids, MI: Eerdmans, 2002), 16.

46. For a summary of such survey data, see Karl W. Giberson and Donald A. Yerxa, *Species of Origins: America's Search for a Creation Story* (Lanham, MD: Roman & Littlefield, 2002), 53–57. Regarding the religious beliefs of American scientists, see Edward J. Larson and Larry Witham, "Scientists Are Still Keeping the Faith," *Nature,* 386 (1997), 435.

47. John Paul II, "Message to the Pontifical Academy of Sciences," *Quarterly Review of Biology,* 72 (1997), 382–83 [emphasis in original].

48. Michael Arnold, *Natural Hybridization and Evolution* (New York: Oxford University Press, 1977), 182–85.

49. Lynn Margulis and Dorion Sagan, *Acquiring Genomes: A Theory of the Origins of Species* (New York: Basic, 2002), 71–77.
50. John Paul II, "Message to Academy," 382.
51. For example, Ernst Mayr, *This Is Biology: The Science of the Living World* (Cambridge: Harvard University Press, 1997), 178.
52. Charles Darwin, *On the Origin of Species,* 1st ed. (Cambridge: Harvard University Press, 1964 fasc. rpt.), 490.
53. Wilson, *On Human Nature,* 192.
54. Stephen Jay Gould, *Wonderful Life: The Burgess Shale and the Nature of History* (New York: Norton. 1989), 14.

GUIDE TO FURTHER READING

The best place to begin reading more about the theory of evolution is with the writings of Charles Darwin. He was not only a profoundly important scientist but also an excellent (and popular) writer. Two of his books, *Journal of Researches* (commonly reprinted under the title *Voyage of the Beagle*) and *On the Origin of Species,* are literary masterpieces. Both books are readily available from several different publishers and in many different editions. Readers may then want to know more about Darwin. Many excellent biographies of Darwin exist. Two recent ones stand out: Janet Browne's two-volume *Charles Darwin* (1995 and 2002) and *Darwin* by Adrian Desmond and James Moore (1991).

Any number of classic monographs deal with the history of science leading up to Darwin. Perhaps the most comprehensive and popular book covering this subject is John C. Greene's *The Death of Adam: Evolution and Its Impact on Western Thought* (1959). Also popular in its day was Loren C. Eisely's *Darwin's Century: Evolution and the Men Who Discovered It* (1958). The standard work dealing specifically with pre-Darwinian developments in geology is Charles C. Gillispie's *Genesis and Geology: A Study in the Relations of Scientific Thought, Natural Theology, and Social Opinion in Great Britain, 1790–1850* (1959). The American story is covered in Herbert Hovenkamp's *Science and Religion in America, 1800–1860* (1978). Martin J. S. Rudwick has written many important books and articles on nineteenth-century geology and paleontology, most notably *The Meaning of Fossils: Episodes in the History of Paleontology,* 2nd ed. (1976) and *The Great Devonian Controversy: The Shaping of Scientific Knowledge Among Gentlemanly Specialists* (1985). Other key books addressing this topic are Peter J. Bowler's *Fossils and Progress: Paleontology and the Idea of Progressive Evolution in the Nineteenth Century* (1976); Roy Porter's *The Making of Geology: Earth Sciences in Britain, 1660–1815* (1977); Stephen Jay Gould's *Time's Arrow, Time's Cycle: Myth and Metaphor in the Discovery of Geological Time* (1987); and J.M.I. Klaver's *Geology and Religious Sentiment: The Effect of Geological Discoveries on English Society and Literature Between 1829 and 1859* (1997). The basic biography of Georges Cuvier is William Coleman's *Georges Cuvier Zoologist: A Study in the History of Evolutionary Thought* (1964). Among the many other fine books on early-nineteenth-century French biology is Toby A. Appel's *Cuvier-Geoffroy Debate: French Biology in the Decades Before Darwin* (1987). On pre-Darwinian evolutionism in Britain, see James A. Secord's insightful *Victorian Sensation: The Extraordinary Publication, Reception, and Secret Authorship of Vestiges of the Natural History of Creation* (2000).

A great and growing body of scholarship examines the reception of Darwinism during the late nineteenth century. Among the numerous biographies of Darwin's collaborators are Adrian J. Desmond's two-volume *Huxley* (1994 and 1997), Peter Raby's *Alfred Russel Wallace: A Life* (2001), and A. Hunter Dupree's *Asa Gray, 1810–1888* (1959). Major works on religious aspects of scientific responses to evolution theory include James R. Moore's, *The Post-Darwinian Controversies: A Study of the Protestant Struggle to Come to Terms with Darwin in Great Britain and America, 1870–1900* (1979), David N. Livingstone's *Darwin's Forgotten Defenders: The Encounter Between Evangelical Theology and Evolutionary Thought* (1987), and Jon H. Roberts's *Darwinism and the Divine in America: Protestant Intellectuals and Organic Evolution, 1859–1900* (1988). Peter J. Bowler's *The Non-Darwinian Revolution: Reinterpreting a Historical Myth* (1988) provides critical insight into how scientists received the theory of evolution. Ronald L. Numbers focuses on the American side of the story in *Darwinism Comes to America* (1998). Edited works dealing with the international reception of evolutionary theory include Thomas F. Glick, ed., *The Comparative Reception of Darwinism* (1988); Thomas F. Glick, et al., eds, *The Reception of Darwinism in the Iberian World: Spain, Spanish America, and Brazil* (2001); and Ronald L. Numbers and John Stenhouse, eds., *Disseminating Darwinism: The Role of Place, Race, Religion, and Gender* (1999). The role of biogeography in late-nineteenth-century evolutionary thought is addressed in two delightful books, David Quammen's *The Song of the Dodo: Island Biogeography in an Age of Extinction* (1996) and E. Janet Browne's *The Secular Ark: Studies in the History of Biogeography* (1983). My book *Evolution's Workshop: God and Science in the Galápagos Islands* (2001) explores the role played by Galápagos research in post-Darwinian debates over the theory of evolution.

A wide variety of books narrate the ever-popular history of the ongoing search for human antiquity and hominid fossils. Participants in this search have frequently contributed highly readable books on the subject, beginning with T. H. Huxley's *Man's Place in Nature* (which remains in print) and including Raymond A. Dart's *Adventures with the Missing Link* (1959); L.S.B. Leakey's *By the Evidence: Memoirs, 1932–1951* (1974); Mary D. Leakey's *Disclosing the Past: An Autobiography* (1984); Richard E. Leakey's *One Life: An Autobiography* (1983); and Donald C. Johanson and Maitland A. Edey's *Lucy: The Beginnings of Humankind* (1981). Pat Shipman relates the discovery of Java Man in *The Man Who Found the Missing Link: Eugène Dubois's Thirty-Year Struggle to Prove Darwin Right* (2001).

The finest history of modern genetics by a key actor in that story is A. H. Sturtevant's *A History of Genetics* (1965). Notable books on this topic by historians of science include Garland E. Allen's *Life Sciences in the Twentieth Century* (1975); Robert Olby's *Origins of Mendelism,* 2nd ed. (1985); Peter J. Bowler's *The Mendelian Revolution: The Emergence of Hereditarian Concepts in Modern Science and Society* (1989); and Jonathan Harwood's *Styles of Scientific Thought: The German Genetics Community, 1900–1933* (1993). The basic biography of Gregor Mendel remains Hugo Iltis's *Life of Mendel,* 2nd ed. (1966). Another important biography in this field is Garland E. Allen's *Thomas Hunt Morgan: The Man and His Science* (1978). Three different perspectives on Francis Galton appear in D. W. Forrest's *Francis Galton: The Life and Work of a Victorian Genius* (1974); Ruth Schwartz Cowan's *Sir Francis Galton and the Study of Heredity in the Nineteenth Century* (1985); and Nicholas Wright Gillham's *A Life of Sir Francis Galton: From African Explorer to the Birth of Eugenics* (2001).

Social Darwinism and eugenics remain active areas of historical scholarship. The starting point to study this episode in American history is Richard Hofstadter's classic *Social Darwinism in American Thought,* rev. ed. (1955). For European developments, see Greta Jones's *Social Darwinism and English Thought: The Interaction Between Biological and Social Theory* (1980); Mike Hawkins's *Social Darwinism in European and American Thought, 1860–1945: Nature as Model and Nature as Threat* (1997); and Richard Weikart's *Social Darwinism: Evolution in German Socialist Thought from Marx to Bernstein* (1999). Diane B. Paul provides a useful introduction to the history of eugenics in *Controlling Human Heredity: 1865 to the Present* (1995). The fundamental work in the field is Daniel J. Kevles's *In the Name of Eugenics: Genetics and the Uses of Human Heredity* (1985). Paleontologist Stephen Jay Gould adds his critical take on the history of eugenics and hereditarian biology in *The Mismeasure of Man*, rev. ed. (1996)—a wonderful read. The sordid tale of German eugenics is told in Paul Weindling's *Health, Race, and German Politics Between National Unification and Nazism, 1870–1945* (1989). Among the leading books on British eugenics are G. R. Searle's *Eugenics and Politics in Britain, 1900–1914* (1976) and Lyndsay Andrew Farrall's *The Origins and Growth of the English Eugenics Movement, 1865–1925* (1985). I examine the practical politics of eugenics in my regional study, *Sex, Race and Science: Eugenics in the Deep South* (1995). Edwin Black adds a well-documented, highly sensational exposé of eugenics in the United States and Germany in *War Against the Weak: Eugenics and America's Campaign to Create a Master Race* (2003).

Opposition to the theory of evolution hardened among many conservative Christians in America during the twentieth century. George M. Marsden provides background for this development in *Fundamentalism and American Culture: The Shaping of Twentieth-Century Evangelicalism, 1870–1925* (1980). The leading study of creationist scientific activity is Ronald L. Numbers's meticulously researched *The Creationists: The Evolution of Scientific Creationism* (1992). The founder of the creation-science movement, Henry M. Morris, provides his own clear view of its history in *History of Modern Creationism* (1984). I relate the history of the Scopes trial in *Summer for the Gods: The Scopes Trial and America's Continuing Debate Over Science and Religion* (1997) and of the ongoing debate over teaching evolution in American public schools in *Trial and Error: The American Controversy Over Creation and Evolution*, 3rd ed. (2003). For the Scopes trial's place in American cultural history, see Paul K. Conkin's *When All the Gods Trembled: Darwinism, Scopes, and American Intellectuals* (1998). The sociology of creation science is discussed in Dorothy Nelkin's *The Creation*

Controversy: Science or Scripture in the Schools (1982) and Christopher P. Toumey's *God's Own Scientists: Creationists in a Secular World* (1994). The creation-science and intelligent-design movements are subject to balanced analysis in Karl W. Giberson and Donald A. Yerxa's *Species of Origins: America's Search for a Creation Story* (2002) and Larry A. Witham's *Where Darwin Meets the Bible: Creationists and Evolutionists in America* (2002). For the exchange over Intelligent Design, compare Phillip E. Johnson's *Darwin on Trial* (1991) and Michael J. Behe's *Darwin's Black Box: The Biochemical Challenge to Evolution* (1996) with Kenneth R. Miller's *Finding Darwin's God: A Scientist's Search for Common Ground Between God and Evolution* (1999) and Robert T. Pennock's *Tower of Babel: The Evidence Against the New Creationism* (1999).

Compared with other chapters in the history of evolutionary science, there is a dearth of scholarship on the history of population genetics, the modern synthesis, and sociobiology. Perhaps the finest book encompassing all these developments is Michael Ruse's comprehensive, thesis-driven *Monad to Man: The Concept of Progress in Evolutionary Biology* (1996). Significant historical studies of population genetics and the modern synthesis include Ernst Mayr's monumental *The Growth of Biological Thought: Diversity, Evolution, and Inheritance* (1982); William B. Provine's broadly conceived biography, *Sewall Wright and Evolutionary Biology* (1986), as well as his briefer *The Origins of Theoretical Population Genetics* (1971); and Vassiliki Betty Smocovitis's insightful *Unifying Biology: The Evolutionary Synthesis and Evolutionary Biology* (1996). Mayr and Provine combined to edit a superb collection of key scientific papers in *The Evolutionary Synthesis: Perspectives on the Unification of Biology* (1980). Joan Fisher Box offers a keen, personal study of her father's work and character in *R. A. Fisher: The Life of a Scientist* (1978). To read further about developments within evolutionary biology and sociobiology since the modern synthesis, the best books are by the biologists themselves. Good choices are Edward O. Wilson's *On Human Nature* (1978) and his autobiography, *Naturalist* (1994); Steven Jay Gould's *Wonderful Life: The Burgess Shale and the Nature of History* (1989); Richard Dawkins's *The Selfish Gene* (1976); and Niles Eldredge's *Reinventing Darwin: The Great Debate at the High Table of Evolutionary Thought* (1995).

Various books present the broad sweep of the history of evolutionary science. Peter J. Bowler's *Evolution: The History of an Idea* (1984) provides the most complete summary of scientific developments in the field. Michael Ruse's *The Evolution Wars: A Guide to the Debates* (2000) relates this history in an accessible textbook format, with documents. Philip Appleman has assembled an exceptional array of primary documents relating to Darwin and Darwinism though time in his tightly edited *Darwin: Texts and Commentary*, 3rd ed. (2001). Other notable anthologies include *The Portable Darwin* (1993), edited by Duncan M. Porter and Peter W. Graham, and *The Darwin Reader*, 2nd ed. (1996), edited by Mark Ridley. The Teaching Company offers my own twelve-lecture course, *The Theory of Evolution: A History of Controversy* (2002), on audio tape, compact disk, and video.

INDEX

About the Author

EDWARD J. LARSON is Russell Professor of History and Talmadge Professor of Law at the University of Georgia. He is the recipient of multiple awards for teaching and writing, including the 1998 Pulitzer Prize in History for his book *Summer for the Gods: The Scopes Trial and America's Continuing Debate Over Science and Religion.* His most recent book is *Evolution's Workshop: God and Science on the Galápagos Islands.* His articles have appeared in dozens of journals, including *The Atlantic Monthly, Nature, The Nation,* and *Scientific American.*

A Note on the Type

The principal text of this Modern Library edition
was set in a digitized version of Janson, a typeface that
dates from about 1690 and was cut by Nicholas Kis,
a Hungarian working in Amsterdam. The original matrices have
survived and are held by the Stempel foundry in Germany.
Hermann Zapf redesigned some of the weights and sizes for
Stempel, basing his revisions on the original design.